PURSUIT OF GENIUS

PURSUIT OF GENIUS

Flexner, Einstein, and the Early Faculty
at the Institute for Advanced Study

Steve Batterson

Emory University

A K Peters, Ltd.
Wellesley, Massachusetts

Editorial, Sales, and Customer Service Office

A K Peters, Ltd.
888 Worcester Street, Suite 230
Wellesley, MA 02482
www.akpeters.com

Library of Congress Cataloging-in-Publication Data

Batterson, Steve, 1950–
  Pursuit of genius : Flexner, Einstein, and the early faculty at the Institute for Advanced Study / Steve Batterson.
    p. cm.
  Includes bibliographical references and index.
  ISBN 13: 978-1-56881-259-5 (alk. paper)
  ISBN 10: 1-56881-259-0 (alk. paper)
   1. Mathematics–Study and teaching (Higher)–New Jersey–Princeton–History. 2. Institute for Advanced Study (Princeton, N.J.). School of Mathematics–History. 3. Institute for Advanced Study (Princeton, N.J.). School of Mathematics–Faculty. I Title.

QA13.5.N383 I583 2006
510.7'0749652--dc22

                                                          2005057416

    [Cover acknowledgments here]

Printed in India
10 09 08 07 06   10 9 8 7 6 5 4 3 2 1

In memory of

Deane Montgomery and Armand Borel,

two professors from the second generation of IAS mathematics faculty. In our interactions I was inspired by the generosity of Deane's support and Armand's scholarship.

CONTENTS

PREFACE   ix

1   ABRAHAM FLEXNER, DANIEL COIT GILMAN, AND
    THE JOHNS HOPKINS TEMPLATE   1

2   THE ROCKEFELLER MODELS   15

3   FUNDING DREAMS   31

4   DECIDING WHERE TO START   55

5   THE FIRST HIRES   81

6   FLEXNER AND VEBLEN BUILD THE SCHOOL
    OF MATHEMATICS   119

7   LAUNCHING THE OTHER SCHOOLS   159

8   COMPETING FOR RESOURCES   189

9   MOUNTING A COUP   215

10 FAST FORWARD   239

SOURCES AND ACKNOWLEDGMENTS   265

INDEX   293

Abraham Flexner, 1923. (Courtesy of the Rockefeller University Archives.)

PREFACE

It was 1980 and I was about to join a remarkable intellectual community. Four years earlier I had received my PhD in mathematics and begun an academic career. Now I was going to the Institute. For other mathematicians the phrase "going to the Institute" said it all. Friends and relatives required additional information. "I am going to a place in Princeton, New Jersey, called the Institute for Advanced Study." When this evoked a blank look, as it normally did, I might add, "It is where Einstein worked."

Just as Einstein is the icon for scientific genius, the Institute holds a special symbolism for mathematicians. The Institute for Advanced Study began in 1930 through the vision of Abraham Flexner. Flexner was a figure of considerable influence during the first half of the twentieth century. He made his mark in 1910 with a scathing exposé of the deficiencies in American medical education. Flexner's revelations called for drastic action. Over a decade-long period he served as the architect of a Rockefeller philanthropic initiative that dramatically upgraded American medical schools.

When Flexner retired from the Rockefeller Foundation, it was with the satisfaction that his career had been essential to the modernization of American medicine. Still, he had a distinctly different ambition that remained unfulfilled. As a long-time observer of higher education, Flexner was convinced that the United States should possess an exclusively graduate university with an ideal environment for research. There, a small faculty of geniuses would direct the studies of a few disciples while pursuing their own discoveries.

With the power to direct millions of dollars to selected universities and hospitals, Flexner had accumulated a stunning collection of contacts among academic, business, medical, and political leaders. When department store magnate Louis Bamberger and his sister Carrie Fuld began seeking advice on devoting their fortune to the creation of a new medical school, it was inevitable that their consultations would lead them to Abraham Flexner. Out of these discussions Bamberger and Fuld decided to endow a graduate university with the 63-year-old Flexner as director.

Flexner was empowered to select the initial personnel and areas of study. He bypassed such intellectual staples as literature, philosophy, and classics. He demurred as well on economics, physics, and other directly applicable subjects. Among all conceivable choices, Flexner elected to begin with a School of Mathematics. Mathematics was a curious place to start. With its abstract nature, higher mathematics is a subject that is fundamentally intellectual. Yet, its discoveries are normally inaccessible to the larger intellectual community.

Flexner himself possessed no knowledge of mathematics.[1] Despite this limitation, he was extraordinarily successful in his first faculty hires. They were the mathematician Oswald Veblen and physicist Albert Einstein. Einstein and Veblen were soon joined by John von Neumann and three other stars from the mathematical powerhouses of Princeton, Göttingen, and Harvard. Suddenly the Institute roster was the strongest of any mathematics department in the world.

The graduate study aspect never materialized. From the beginning the faculty identified and hosted scholars who had already received their doctoral degrees. These visitors, who became known as members, typically remained at the Institute for a year. Fifty years later I was a member and marveling at Flexner's legacy.

I unpacked some clothes at my furnished apartment on the edge of the Institute grounds. Walking a few minutes down Einstein Drive I reached my office in the building originally constructed to house von Neumann's electronic computer project. There were neither classes to teach nor service duties to perform. It was an environment totally devoted to the enhancement of research.

The School of Mathematics was not only the oldest, but also the largest of the Institute's four schools. It had about 60 members and seven faculty. Among the faculty were three Fields Medalists and another person who would soon be awarded this so-called Nobel Prize of mathematics. Colloquia and ad hoc seminars were abundant. However, every member had complete liberty to follow any schedule and pursue any topic. When an office became confining, a walk along the magnificent trails through the Institute's woods offered a stimulating diversion. Lunch was prepared by a first-class European chef, and tea and cookies were available every afternoon at three.

The chef was a recent addition to the staff, and we were told that the food had not always been so good. More substantive aspects of the history of this mathematical paradise remained mysterious. Flexner himself once wrote, "Institutions like nations are perhaps happiest if they have no history."[2] He did little to violate this principle with the two chapters that dealt with the Institute in his autobiography. "The Institute for Advanced Study" and "Finding Men" provide a rosy account of the founding and the recruitment of faculty. From a factual perspective Flexner's narrative is largely accurate, but there are significant omissions that leave a distorted picture in portraying the challenges he faced. The internal issues involved in governing the Institute, and Flexner's reactions, will be developed in this book. The external historical events are well known, even if their relevance to the origins of the Institute for Advanced Study is not.

Flexner's tenure as Institute Director occurred during the 1930s, a decade now inextricably associated with the Great Depression and the ascendance of Adolf Hitler. It must be remembered, however, that these events unfolded in an unprecedented fashion and their magnitude, in the first few years, was difficult to comprehend. Indeed, one chilling aspect of research into the history of this period is the cavalier manner in which the public initially regarded Nazism and the Depression, oblivious that they were witnesses to catastrophic events of the century. Early in 1932 one future Institute professor wrote to Veblen from Göttingen,

> I am a little bit more at ease concerning the political future of Germany. I now have the impression that Hitler is lacking the courage to take responsibility for carrying out his radical ideas and that he is happy if he can stay in the opposition. If National Socialism will be established then probably only in a very much watered down form which will be at most economically dangerous.[3]

It would take another full year for this professor and his Jewish wife to appreciate the hopelessness of their future in Germany. Meanwhile, the economic crisis initially aided Flexner in his recruitment of faculty, but it eventually contributed to his own downfall.

It is curious that, although its founding was a seminal event in intellectual history, the Institute's existence is unknown to many scholars. This book examines the creation of the Institute and its early evolution. The emphasis is on the first generation of mathematics faculty. Archival correspondence provides insight into Flexner's thinking as he decided to begin with mathematics. Of paramount importance was his perception of the world's leading scholars and their availability. The recruitment of mathematics faculty tested Flexner as he persevered through a number of tribulations to assemble his team. Veblen led this group as they established an international seat of mathematics and shifted the Institute's focus away from graduate study. Programs in other subjects developed under markedly different circumstances. Today the Institute maintains its mystique.

President Gilman and early Johns Hopkins faculty in 1901: from left to right, seated, Gildersleeve, Gilman, Remsen; standing, Rowland, Welch. (Courtesy of Ferdinand Hamburger Archives of The Johns Hopkins University.)

CHAPTER ONE

ABRAHAM FLEXNER, DANIEL COIT GILMAN AND THE JOHNS HOPKINS TEMPLATE

People have long been fascinated by the topic of who is best at various endeavors. A striking example is the interest generated by the 1992 United States Olympic basketball team that included Michael Jordan, Magic Johnson, and Larry Bird. Regarded as the greatest collection of basketball talent ever to perform together, the group's dazzling athleticism and skill led to its designation as the Dream Team. Super Bowls and World Cups are spectacles whose champions are the immediate focus of worldwide adulation. In intellectual pursuits, superlative standing is evaluated more subjectively. At the close of the nineteenth century the universities of Paris and Berlin attracted many of the world's outstanding scholars. It was the German university town of Göttingen, however, that was regarded as the greatest center for mathematical research. At that time the star of the Göttingen faculty was David Hilbert, who was following Carl Friedrich Gauss and Bernard Riemann in an extraordinary mathematical heritage. The pilgrimage to Göttingen was imperative for serious mathematicians in Europe and the United States.

In 1933 the New Jersey community of Princeton abruptly replaced Göttingen as the world's leading center for mathematical research. This sudden transfer of mathematical prestige was a consequence of the diverse aspirations of two laymen: Adolf Hitler and Abraham Flexner. Just as Hitler's policies were eviscerating German universities, Flexner opened the Institute for Advanced Study with its School of Mathematics faculty that included Albert Einstein and John von Neumann. Flexner's achievement was remarkable. Operating in the midst of the Great Depression, he created an unprecedented environment for mathematical research and succeeded in assembling the greatest mathematics department in the world. Flexner was an unlikely patron of the subject. His own background was totally devoid of mathematical study.

Abraham Flexner was born November 13, 1866, in Louisville, Kentucky. He was known as Abe to family and friends.* Abe's parents were Jewish immigrants from Europe. The family business enterprises rose and fell with the economic tide of the times, crashing in the Civil War and again in the Panic of 1873. As the sixth son among the nine children, Abe was permitted to complete school while his older siblings dropped out and went to work. Through the generosity of his brother Jacob, Abe became the first family member to attend college.[1]

For the remainder of his life Abe felt indebted to Jacob. Abe was grateful not just for the act of familial sacrifice but even more so for Jacob's selection of Johns Hopkins University. Although the university was then in its infancy, it was breaking new ground in American higher education. Abe felt privileged to have arrived at just the right time and place. The Johns Hopkins experience profoundly shaped Flexner's values. In subsequent years he looked on with dismay as American universities followed other models. The Institute for Advanced Study arose out of Flexner's desire to resurrect the founding ideals of Johns Hopkins University.[2]

The Baltimore financier Johns Hopkins had died a decade before Flexner's enrollment, leaving a $7 million endowment to split between the founding of a university and a hospital. Hopkins designated two overlapping groups of prominent citizens to serve as trustees for the institutions. The university trustees found themselves in charge of a record-breaking bequest, with few attached stipulations. Proceeding cautiously, they consulted with the presidents of Harvard, Yale, Cornell, and Michigan. The organizational task was no doubt simplified when the experts independently, but unanimously, recommended Daniel Coit Gilman to lead the new institution. Gilman arrived in Baltimore just prior to the 1875 New Year to meet with the trustees. It was clear that if he made a satisfactory impression, then Gilman would be offered the presidency.[3]

As Flexner later wrote in his book on Gilman, "Thus, in his forty-fourth year—the very prime of his life—Gilman's great opportunity came to him and he was ready for it. He had been preparing for it since his early twenties."[4] Gilman was from a prosperous Connecticut family. He excelled as an un-

---

*In the following narrative Abraham Flexner is referred to as Abe in passages that include mention of other members of the Flexner family.

dergraduate at Yale in the mid-nineteenth century. For an ambitious, cultivated young man, a variety of career opportunities were available. Gilman,
however, wanted further experience and preparation prior to settling on a
vocation. Unfortunately, his academic options were somewhat limited in a
country where no PhD program existed.

Following graduation, Gilman briefly continued his studies at Harvard
and Yale. He soon realized that it was necessary to go abroad for the experience he was seeking. Gilman then embarked on two years of work and travel
in Europe. In St. Petersburg and Paris he obtained minor foreign service positions. Everywhere he absorbed the culture and met new people. The Berlin
stop was notable for the connections he established with academic scholars.
Although Gilman rejected the notion of pursuing a PhD, he keenly observed
the diverse local educational systems. When he returned to America in 1855,
Gilman contemplated possible careers in religion and education.

While Gilman would dabble in preaching, his interest in education exerted a stronger pull on his career. Over the next seventeen years, Gilman
served Yale in a variety of capacities. Many, but not all, were associated with
the Scientific School, a separate branch of the university. Yale was typical of
American universities of the time in placing science in a position subordinate
to the humanities. Gilman began as a fundraiser for the Scientific School
and then moved into other administrative positions. Throughout his tenure
Gilman proselytized for the elevation of science to the standing it enjoyed in
European schools. Along the way he became university librarian and professor of geography. Somehow Gilman also found time to maintain an active
role in public education, serving on local and state boards.

Gilman was a forceful advocate who used his administrative positions
to push for reforms. As librarian he reallocated funds to raise science and
literary accessions to the level of theology accessions. Often his campaigns led
to disagreements with superiors over the priorities for limited funds. Despite
these conflicts Gilman earned recognition as an educational leader. Given
his ambition to define new directions in higher education, it might seem
surprising that Gilman declined offers of the presidency at the Universities of
Wisconsin and California in 1867 and 1870. It is likely that his aspirations
were to rise to that position at his alma mater. When a more conservative
colleague was selected for the role, Gilman accepted a renewed offer from
California in 1872.

There is another possible explanation for Gilman's original refusal at California. He may have foreseen the opposition that his initiatives were to en-

counter from regents, politicians, and newspapers. Being University of California president did not provide a free hand to implement new policies. It merely shifted the struggle to a higher level. Within two years a frustrated Gilman tendered his resignation. The regents persuaded him to remain, but Gilman recognized that their reconciliation was likely to be short-lived. He was ready to discuss an offer from another university.

These were the circumstances when the opportunity arose in Baltimore. Gilman had an agenda for university reform and Hopkins offered a clean slate endowed with a substantial budget. From his bitter California experience, Gilman knew that an aggressive program needed the full backing of the trustees. When Gilman arrived for the Hopkins interview, he pulled no punches in presenting a bold vision for a university dedicated to research and graduate education. Professors were to be scholars of demonstrated accomplishment. They were to be well compensated and given sufficiently low teaching loads to permit continued research and supervision of advanced students.

Aside from the exclusion of undergraduate education, the picture painted by Gilman was much the same as would be presented today by a presidential candidate at a quality university. The 1875 American landscape, however, was vastly different. American universities provided little incentive for their faculty to engage in scholarly work. High teaching loads and other duties left little time. It was the European universities that promoted research, especially those in Germany. To receive advanced training Americans were forced to go abroad. Consider that the first United States PhD was awarded in 1861 and the first in mathematics one year later, both at Yale. It would be eleven more years before Harvard made these respective doctoral conferrals. Graduate education was just emerging in the United States, and then only at long established Ivy League schools. Gilman was proposing to open Hopkins as a purely graduate institution, leaving undergraduate education to the currently existing colleges. There was no precedent in the United States. By all accounts Gilman was an effective advocate. Despite the revolutionary nature of his plan, he was quickly offered the job.[5]

The Baltimore press was less enchanted with Gilman's vision, tarring it as elitist and of little use to the community. Over the next half year Gilman and the trustees thrashed out an organizational plan. To assuage his critics, Gilman made the concession to incorporate both graduate and undergraduate programs in the model. On other matters there was full agreement. This was fortunate because, with the opening scheduled for fall 1876, it was important to establish priorities. It was decided to hold off on the expense of constructing a new campus. Modest temporary quarters would suffice and permit some accumulation of investment dividends. As for areas of study, medicine would be deferred until the hospital were completed. The initial focus was to be on arts and sciences, especially at the graduate level. A salary range of $3,000 to $5,000 was adopted for professors, offering a higher ceiling than either Harvard or Yale. The maxim "men over buildings" came to be associated with Gilman.

To succeed, Gilman needed to find men* for his professorships who were both dedicated to original research and capable of inspiring advanced students to pursue the same course. In 1875 there were few such men in the United States. Gilman was in no position to quibble over subject area. He adopted a policy that he coined as "good work in a limited field."[6] The principle has been aptly characterized as follows: "If initially he could secure a Latinist but no chemist who met his standards, then Hopkins would have a Department of Classics but no Department of Chemistry."[7] Gilman made trips through the United States and Europe to recruit and to promote the new university. When Hopkins opened in 1876 there were six professors, one each in mathematics, Latin, Greek, physics, chemistry, and biology. The heavy representation of science was unusual for an American university at that time.

The choice of the first mathematics professor illustrates Gilman's approach and courage. James Joseph Sylvester was an accomplished, but unemployed, British algebraist in his early sixties. Although Sylvester was barred from most university positions in England because he was Jewish, he had not succeeded in the few teaching opportunities that had been available to him. His previous professorships in the United States and England had ended under doubtful circumstances. Still he was a world class mathematician with strong support from respected scientists. Gilman weighed the many positives and negatives as he met with the colorful Sylvester in London. Gilman decided that Sylvester was the man he needed to launch mathematics at Hopkins. An offer was made and negotiations were tough. After a variety of ultimatums, counteroffers, and refusals and the intervention of an intermediary, an agreement was reached for Sylvester to receive an above-scale salary of $6,000.[8]

---

*"Men" was the term employed by Gilman. While its generic use in this context did not always completely exclude women, Gilman's search was confined to a single gender.

Gilman's judgment of Sylvester was correct. Eight students received PhDs under Sylvester's direction during his seven and a half years at Hopkins. While this would have been an impressive yield at any time, it was greater than the entire number of mathematics PhDs awarded on United States soil prior to Sylvester's arrival. Significant mathematics graduate education in America began at Hopkins with Sylvester.

Another triumph of Gilman's talent search was Henry Rowland. Rowland was a young physics instructor at Rensselaer when he was discovered by Gilman one year prior to the opening of Hopkins. Rowland was quickly placed on the Hopkins payroll and given time to devote to research and European consultation. During that year Rowland made fundamental discoveries in electricity and went on to a distinguished career in research, if not in teaching. In Greek the choice was Basil Gildersleeve, a middle-aged Confederate veteran of the Civil War. Gildersleeve established Hopkins as an important center for classical studies. As diverse as their backgrounds and ages were, the common element was Gilman's ability to evaluate men. It was a quality that Flexner would admire greatly and see in himself.

In 1884 seventeen-year-old Abe Flexner arrived from Louisville to enroll in the heady environment of Johns Hopkins. It began an experience that shaped his life. Short of funds, Flexner overloaded on courses and took exams to exempt out of subjects he had already studied. Two years later he left Hopkins, having gained some knowledge of the classics and an undergraduate degree. More important to the future of the Institute for Advanced Study was that Flexner found in President Gilman a hero that he aspired to emulate.

In his autobiography Flexner reflected glowingly on his time at Hopkins, "Research was in the air we breathed."[9] Woodrow Wilson was a graduate student whom Flexner often observed working in the library. There were no extracurricular activities and Flexner spent all of his time studying. Administration was minimal and efficient. When Flexner needed to deal with an exam conflict, he simply spoke to Gilman, who handled it appropriately. While Flexner's memories of Hopkins, like so many other events, were no doubt embellished in his autobiography, his veneration of Gilman was not exaggerated. Flexner was convinced that Daniel Coit Gilman was the greatest figure in the history of American higher education.

Following his graduation from Hopkins in 1886, Flexner returned to Louisville. Now it was his turn to support his mother and siblings. Abe began teaching in the high school he had attended just two years earlier. In addition to classes on Latin and Greek, he supplemented his income by tutoring. By 1890 there was sufficient demand for Abe's personalized skills that he was able to open his own small school. Flexner's specialty was college preparation for underachieving boys from well-to-do families. The school grew and flourished, remaining part of the Louisville scene for fifteen years.

Family was always foremost in Abe's mind. Having established a stable income from his teaching, Abe offered to finance a year of education at Hopkins for his brother Simon. What was a laudable attempt to repay his own opportunity would turn into a surprising investment in medical science. Simon was a late bloomer. He dropped out of school in the middle of sixth grade, and went on to demonstrate neither talent nor ambition in a variety of unskilled jobs. The turning point of his life followed a near fatal bout with typhoid fever in his late teens.[10]

Simon was then apprenticed to an area pharmacist. With the perspective gained from his near-death experience, an able student and worker suddenly emerged. Following his certification, Simon began to practice in his brother Jacob's pharmacy. For the first time in his life Simon showed initiative and intellectual curiosity. He participated in discussions with the local physicians who gathered in the store. Most significantly, Simon began to explore a new world through the pharmacy's microscope. He became a self-taught student of pathology, and his skills were acknowledged among the Louisville medical fraternity.

As his fascination with the subject grew, Simon thought of opening his own pathology laboratory in Louisville. At that time it was quite an idea. There were no pathology labs in Louisville, nor was Simon aware of the existence of one elsewhere. He had taken the creative step of conceiving a career around his passion. Local doctors would bring blood, urine, and other specimens for analysis. To pursue this venture would first require some training from an experienced pathologist, but where could this be obtained? In the literature available to Simon it appeared that the development of pathology was restricted to Germany. Abe's timing could not have been better.

Looking ahead to the start of the Hopkins medical school, Gilman had begun to search for men. Traditionally such personnel were recruited locally from practicing physicians. Gilman had higher ambitions and followed the course that he had pursued in identifying the original arts and sciences faculty. He sought advice from leading scholars, and traveled to Europe to survey the field. As had happened with Sylvester in mathematics, one name was promoted by several respected sources. William Welch was an American who had studied pathology in Germany. He was frustrated by the lack of opportunity for research in his position at the Bellevue Hospital Medical College in New York. In 1884 Gilman invited Welch to Baltimore. The offer of a pathology professorship was made and accepted, beginning an association that would launch a new era in American medical education.

Rather than have Welch immediately begin a graduate pathology program, Gilman had the vision to finance a year of further work abroad. Welch returned to Germany, where the infant sciences of pathology and bacteriology were undergoing the most intense study. The first year that Welch was in residence at Hopkins was Abe's last year. During this time Welch set up his lab and gave a series of lectures on "Microorganisms in Disease." To appreciate Welch's timing, consider that these were "the first comprehensive lectures on bacteriology in America."[11] Abe attended some of the talks and appreciated the relevance to the microscopy that was occupying his older brother. It would take a few years for him to gain the financial means, but Abe succeeded in matching the only pathologists that he knew.

Simon arrived at Hopkins in 1890. It was one year after the opening of the Hospital and it would be another few years before financial conditions would permit the start of the Medical School. Welch was conducting research, gathering medical staff, and supervising the training of pathologists. Simon was 26 years old and away from his family for the first extended period of his life. With barely the resources for a year of study, there was little time to ease into the new environment. Simon made the most of it, and his talent was quickly appreciated. One of Gilman's first innovations in the Hopkins graduate program had been to make available graduate fellowships on an unprecedented scale. This financial support had the immediate effect of attracting the most promising graduate students. The fellowship in pathology became available and was awarded to Simon at the end of his first year, providing him essentially independent means to continue his education.

Given his lack of formal training prior to Hopkins, Simon's scientific ascendance was astounding. By his third year he achieved the faculty rank of assistant. There was no longer any thought of returning to Louisville. Flexner was Welch's protegé. He was a vital instructor in the revolutionary program of the new medical school and his research was attracting attention. Two years later he was promoted to associate professor. Offers from a number of medical schools began to arrive, as they did for others trained by the Johns Hopkins medical faculty. Finally, in 1899 Simon Flexner accepted a professorship of pathology at the University of Pennsylvania, the oldest medical school in the country.

When Gilman established Johns Hopkins University his mission was to produce stories such as Simon's. An American. receives advanced training from the senior staff in cutting-edge research techniques and then exerts an impact with his own scholarship. There were many other examples in the medical arena. A contemporary student of Simon was Walter Reed, whose later field work established that yellow fever was transmitted by mosquitos. Other young faculty members were called to professorships at Harvard and Pennsylvania.

Welch was the perfect choice to implement the medical phase of Gilman's program. He followed the university's founding principles in linking research with medical instruction. Moreover, Welch recruited scientists in touch with the current European breakthroughs. In the late nineteenth century a scientific revolution was underway. Advances in European laboratories were revealing the pervasive role of microorganisms in causing diseases. As the Hopkins medical school was opening, a Berlin scientist made the astonishing discovery that immunity from diphtheria was conferred by injection of its toxin. Unfortunately, such concepts failed to reach the American medical curriculum. Training in these programs was typically given by local doctors who were uneducated in the European developments and skeptical of their efficacy. Not only was Hopkins the first American program to incorporate bacteriology and immunology into its course of study, but it was through its proselytes that these fundamental concepts spread to the United States' medical community. This was a fundamental advance in American education. To Abe Flexner there was ample credit to be shared by Gilman and Welch. Both had the vision to appreciate research and to select men who could train men.

The program in mathematics, although impressive, had less lasting influence. Under Sylvester, it had the distinction of becoming the first substantial mathematics graduate program in the United States. Its most tangible legacy was the founding of the *American Journal of Mathematics*. Intended to address the dearth of opportunity for scientific publication in the United States, the periodical continues today as one of the world's leading mathematics journals. On the down side, Sylvester's doctoral students had little further scholarly impact. Moreover, the success of the mathematics program was largely dependent on Sylvester. When he resigned to return to England in 1883, there was a serious void in leadership. It was at this point that Gilman missed a valuable opportunity.

Sylvester lobbied for the appointment of Felix Klein as his successor. Klein, then 34 years old and at Leipzig, was one of the foremost mathemati-

cians in the world. Gilman was a great admirer of German scholarship, and
had just returned from the country on his medical recruiting trip. Through
his extensive contacts, Gilman was knowledgeable about the university system. He knew that calls from other institutions were routinely used by German professors to leverage improvement in their current positions. Gilman
was convinced that offers to German faculty were unlikely to succeed in luring
them to Hopkins.

Sylvester was determined to have Klein carry on the mathematics program
that he had begun at Hopkins. To bring this about he applied his persuasive
powers to both Klein and Gilman. An offer was eventually made at a salary
of $5,000, the maximum of any Hopkins professor following the departure
of Sylvester. Klein expressed interest, but insisted on the $6,000 figure that
he knew had been paid to Sylvester. In addition, he requested some form
of pension or security for his family, as was already available to him under
the German system. Gilman was either unable or unwilling to persuade the
trustees to meet these terms or provide a counteroffer. A repetition of the
earlier offer was rejected by Klein.[12]

A number of factors, both at home and abroad, may have contributed to
the administration's reluctance to enter negotiations with Klein. It is possible
that Gilman was aware that Klein had just emerged from a breakdown caused
by overwork. In fact he had already done his best mathematics. Meanwhile
at Hopkins, Sylvester's age and eminence had made him an understandable
exception to the salary ceiling. If the much younger Klein were offered the
same terms, it was likely to cause difficulty among the other professors. While
Hopkins' endowment had seemed ample at its founding, the instability of its
stock portfolio called for frugality. Finally, Gilman was beginning his buildup
for the medical school and was budgeting for additional faculty in this area.
Given Gilman's cynicism about landing a first-class German professor, it is
not surprising that Klein slipped away. This was unfortunate for two reasons.
Klein was seriously interested in coming to Hopkins, and he would have been
a spectacular appointment.

Klein remained at Leipzig for two more years and then moved to Göttingen. With the decline of his research activity came a shift of energy into
teaching. Employing lucid lectures to convey his unusually broad and deep
understanding of mathematics, Klein became a "master teacher."[13] As his remarkable teaching skills became known by word of mouth, both American
and European mathematics students traveled to attend Klein's classes. Many
obtained their PhD under his direction, while others received predoctoral

and postdoctoral training. At the turn of the century the Klein pedigree was a common thread running through American mathematics departments at the front rank in research and graduate training. Klein had accomplished, from Europe, precisely what Gilman had hoped to achieve by his own faculty. Below is a partial list of his students and the universities where they eventually settled and exerted their influence:[14]

Princeton:	Henry Fine (PhD) and
	Henry Thompson (PhD)
Chicago:	Oskar Bolza (PhD) and
	Heinrich Maschke
Harvard:	Maxime Bôcher (PhD) and
	William Osgood
Wisconsin:	Edward Van Vleck (PhD)
Cornell:	Virgil Snyder (PhD)
Northwestern:	Henry White (PhD)
Berkeley:	Mellen Haskell (PhD) and
	Irving Stringham
MIT:	Frederick Woods (PhD) and
	Henry Tyler
Michigan and Columbia:	Frank Cole

Klein's success at training the next generation of scholars distinguished him from his Göttingen predecessors Gauss, Dirichlet, and Riemann. While Klein was unable to match the theorems of these superstars, he introduced a mathematical research culture to Göttingen. Then, in 1894, Klein succeeded in bringing David Hilbert to the university. Although Klein's mathematical skills were in decline, Hilbert was at the top of his game. The double attraction of Klein and Hilbert lifted Göttingen to the highest standing in the mathematical world, a position that it would maintain until 1933.

Hindsight reveals that Gilman made a disastrous personnel move in his failure to secure Klein. Sylvester's chair was filled by Simon Newcomb, a mathematical astronomer who was one of the United States' most influential scientists. Newcomb assumed the professorship on a half-time basis, maintaining his position as superintendent in the Nautical Almanac Office. Sylvester had invested his heart and soul into leading a program in abstract mathematical research. Newcomb was a part-timer whose primary interest was in astronomy. While Newcomb remains an important figure in the development of American science, under him the Hopkins graduate program in mathematics lost the edge that it had established under Sylvester. Over time it was surpassed by Chicago, Harvard, and Princeton.

Despite the missed opportunity with Klein, Gilman's tenure as president must be viewed in the context of American intellectual history. Others who followed would raise the bar, but it was Gilman who first set the bar and established research and graduate education as essential features of American universities. At the Johns Hopkins University twenty-fifth anniversary celebration, President Charles Eliot of Harvard placed this accomplishment in perspective:

> President Gilman, your first achievement here, with the help of your colleagues, your students, and your trustees, has been, to my thinking—and I have good means of observation—the creation of a school of graduate studies, which not only has been in itself a strong and potent school, but which has lifted every other university in the country in its departments of arts and sciences. I want to testify that the graduate school of Harvard University, started feebly in 1870 and 1871, did not thrive until the example of Johns Hopkins forced our Faculty to put their strength into the development of our instruction for graduates. And what was true of Harvard was true of every other university in the land which aspired to create an advanced school of arts and sciences.[15]

Eliot went on to pay tribute to Gilman's role in advancing medical education and basic research. No Hopkins graduate could have been more effectively indoctrinated with this philosophy than Abraham Flexner. It would be many years, however, before he would be in a position to serve as Gilman's disciple. First there was a family to support and debts to repay. Through his Louisville school, Abe succeeded in both aspects. He was able to send Simon to Hopkins and a sister to Bryn Mawr. Moreover, his school remained stable as the Panic of 1893 devastated other family businesses.

Abe had ambitions to achieve some form of intellectual greatness, but a sense of duty tied him to Louisville in the years immediately following the panic. The Flexner family was financially dependent on the school that even provided employment for the sisters. Although Abe was the youngest son, he had taken over from Jacob as the leader of the family. Adding to these responsibilities was his marriage to an aspiring writer, Anne Crawford. By the end of the century the financial situation had improved, but Abe remained in Louisville at his school.

Simon was confounded by the letters that he received from his increasingly unfulfilled brother. Abe was depressed by his inability to escape the tired role as Louisville schoolmaster, yet he seemed unable to take the sort of plunge that had transformed Simon's life. Simon came to believe that Abe was reinforcing his fear of failure with an exaggerated sense of family duty.[16] When Anne's play *Mrs. Wiggs of the Cabbage Patch* became a Broadway success, Abe had an unqualified opportunity to pursue his own creative path. In 1905 he closed the school. Abe would resume his education at Harvard and abroad, but, unlike Simon, a particular calling had yet to reveal itself. Although Abe had experience, opinions, and interest in the field of education, it was not a subject that he viewed as suitable for graduate study.

Flexner's vague plan was to begin by exploring the subjects of psychology and philosophy at Harvard and then proceed to Berlin. Somewhere in these settings, Flexner expected to find his niche. It did not occur at Harvard. After an unsatisfactory year of laboratory work in psychology, Abe was ready to move on to the European phase. His two years in Germany, 1906-8, gave him a glimpse of German culture and education prior to World War I. At that time Germany led the world in much of the research that Abe had come to value while at Hopkins. Secondary education also provided interesting contrasts with that in the United States. While Abe seemed to enjoy his study in Germany more than at Harvard, he remained without a commitment to a path of research. The German experience had expanded his overall view of education and positioned Abe to formulate his own dogma, but it did not make him a scholar.

At the age of 41 Abe had completed his formal education. Even with a master's degree, Abe probably felt that he was a failure. Gilman and Hopkins had prepared and challenged him to push the frontiers of knowledge. Simon had succeeded, but Abe's only claim was his vicarious support of his brother.

Simon Flexner, 1914. (Courtesy of the Rockefeller University Archives.)

- CHAPTER TWO -

THE ROCKEFELLER MODELS

An entrepreneur makes a fortune and provides funds to endow an institution of higher learning. A visionary is selected to lead the project and American scholarship is lifted to a new level. What Daniel Coit Gilman achieved with the opportunity offered by Johns Hopkins was a script that would be repeated with other players. Both Flexner brothers would eventually reprise the Gilman role.

The financial means for these ventures were donated by some of the capitalists who acquired enormous wealth amidst the industrial development of the late nineteenth century. Hopkins' success was in the railroad business. The icon for the self-made millionaire was John D. Rockefeller. Not only did Rockefeller's leadership of Standard Oil offer an effective model for global corporate domination, but his approach to charitable activity opened innovative paths of munificence on a similarly unprecedented scale.

Born in 1839, Rockefeller's early life was spent in western New York. His father, William, made a living as a con man and charlatan. To find dupes for his cancer cures and other schemes William took to the road, neglecting his family over long periods of time. With an absent father, John became devoted to his mother Eliza, who looked in turn to her eldest son as the surrogate family leader. For both John and Eliza the local Baptist church became a source of happiness and purpose. Throughout John D. Rockefeller's long life the church would remain at the center of his existence. A recent biography by Ron Chernow makes a strong case that out of his early Baptist immersion, John became convinced that he had a divine calling to acquire wealth and then distribute it as charity.[1]

When John was thirteen, his father moved the family to the Cleveland area and then withdrew even further from their lives. College was out of the question for a teenage head of the household. At sixteen, John obtained a

15

position as assistant bookkeeper in a commission firm that dealt in produce shipping. After a probationary period Rockefeller's wages were set at $25 per month.

Over the next few years Rockefeller worked conscientiously and managed to double his meager salary. Meanwhile he lived frugally, saving money and always donating a generous portion to charity. Most importantly he was an astute observer of the local business scene and gained experience as his firm coped with the Panic of 1857. The following year Rockefeller joined Maurice Clark to begin a partnership in the commission business.

Taking advantage of Cleveland's strategic position in the developing railroad industry, Clark and Rockefeller prospered. The Civil War brought additional opportunities. Rockefeller's greatest triumphs were in the oil business. It all began in 1863 when Clark and Rockefeller took the decisive step of building an oil refinery in Cleveland, just a few years after oil was struck in western Pennsylvania. Rockefeller conceived innovative practices to cut costs and increase the productivity of his refinery.

Working with a different partner, Rockefeller entered into secret deals with railroads. Through a ruthless, predatory campaign the company bought out their less profitable Cleveland competitors. Soon the Rockefeller partnership was the world's largest refiner. In 1870 Standard Oil Company (Ohio) was incorporated and embarked on a program to monopolize the oil industry. The Panic of 1873 and subsequent depression enabled Standard Oil to skirt the law and covertly take over refineries in other states. Meanwhile they expanded their interests by entering and dominating first the oil tanker and then the pipeline business. By the end of the decade Standard Oil controlled 90% of American refined oil and Rockefeller was a multimillionaire. When he transferred his base to New York in the 1880s, both the company and his personal wealth sustained their phenomenal growth.

Keeping his charitable contributions proportional to income was both a responsibility and challenge for Rockefeller. After all, God had entrusted him with riches to distribute. For Rockefeller this meant allocating the money in a manner that would strengthen the needy, rather than make them dependent. Given the number of appeals and the scale of available funds, the obligation of personal scrutiny became an overwhelming burden.

Education was an avenue in line with Rockefeller's philosophy of assisting self-improvement. He began to consider establishing a large endowment for a higher education initiative. This was new ground and Rockefeller proceeded cautiously. While Johns Hopkins had left the arrangements to his trustees,

Rockefeller was determined to set the initial course before turning the control over to capable hands. What was clear was that there would be a Baptist element to the institution. In the latter 1880s various American cities lined up to present themselves as worthy venues.

One proposal was for a Baptist college in Chicago. There was a strong case for a regional institution to serve the burgeoning westward growth, but the timing was a bit odd. A Baptist University of Chicago had just closed due to financial difficulty. Out of the old university and its surviving relative, the Baptist Union Seminary, a movement emerged to make a fresh start. Rockefeller's responses to the early overtures were not encouraging.

For a few years Rockefeller pondered New York and other locations. A strong feature of the Chicago proposal was its association with William Rainey Harper. Harper was a brilliant young Biblical scholar who, in 1875 at the age of 18, received one of the first PhDs from Yale. With an eye for talent in both business and church matters, Rockefeller had followed Harper's career from the Baptist Union Seminary to his call to a chair at Yale. Rockefeller was enormously impressed by Harper, whose name was being floated in the Chicago pitches as president of the new university.

When the American Baptist Education Society was created in 1888, the Chicago cause gained a powerful advocate. The executive secretary of the new organization was a persuasive young minister named Frederick T. Gates. Gates quickly became convinced of the merits of founding a Baptist university in Chicago. He joined Harper in an effective double-team of Rockefeller.

Once the Chicago venue was selected, Rockefeller shifted his focus to the logistics. His intention was to make a modest start with a budget that was within the income from the initial endowment. Under these circumstances he preserved the future options of withdrawing from a self-sufficient institution or selecting directions for enhancement with additional funds. The buffer of the American Baptist Education Society permitted Rockefeller to feign detachment, but he was accustomed to wielding the power that accompanied his wealth. Of course he would want Harper to serve as president.

Harper had some leverage of his own. If he were to give up an attractive professorship at Yale, he wanted the resources to launch a first-class graduate program elsewhere. Rockefeller's qualms about rising expenses were mollified by Gates, who had become the point man for the project. Gates' proposal was to begin with the undergraduate college, tentatively planning to follow up a few years later with a graduate program. It was the beginning of Gates' enormously influential role in shaping Rockefeller's philanthropy. The next step was Gates' negotiation of a $600,000 Rockefeller pledge on condition of a successful $400,000 matching fund drive. Even after the money was raised, Harper had not yet committed himself to the presidency. At this point Harper managed to coerce an additional $1 million from Rockefeller for the graduate program.[2]

Early in 1891 Harper accepted the presidency of the University of Chicago. Classes were slated to commence in the fall of the following year. Just as Gilman before him, Harper's ambition was to create a university that would be an innovative model for research and higher education. With Johns Hopkins having demonstrated that such an institution could succeed on American soil, the existence problem had been solved. Harper was emboldened to draft a plan for his university, far more detailed than that of Gilman, introducing significant changes in mission and structure. Outreach was to join traditional on-campus study as a vital educational component. To more efficiently utilize infrastructure Harper devised the quarter system calendar. The number of subjects to be carried by a student was reduced, accompanied by an upgrade in the rigor and demands of each course.

Both Gilman and Harper invested their greatest energy in the recruitment of the research faculty. They knew that the ultimate measure of the university hinged on their success in identifying and procuring world class scholars. The influence of Johns Hopkins in lifting American scholarship meant that Harper surveyed a different field than had Gilman one and one half decades earlier. Not only were there more researchers established in the United States, but their talents were becoming appreciated by the leading universities. It was amidst this competitive environment that Harper sought to lure the best to Chicago.

Harper left no ambiguity in the job description for his graduate faculty. Research, rather than teaching, was the primary responsibility. Course loads were to be exclusively graduate and sufficiently light to leave ample time for individual scholarly work. Harper identified the individuals who met his high standards and personally lobbied each to join him in launching the new university. What he found was depressing. Despite his most persuasive pitch, nobody would come to Chicago under the $6,000 salary ceiling set by the trustees. When this was raised to $7,000 Harper reeled in his first two professors, William Hale in Latin and J. Laurence Laughlin in political economy, both from Cornell.

Even with the higher figure Harper was unable to pry a single senior faculty member away from Harvard or Johns Hopkins. Indeed the threads of institutional status and salary pervaded the negotiations. Harper was asking scholars to defer gratification and share his vision that the University of Chicago would become a great university. Certainly the capitalization was inadequate to sustain such a venture. Harper's approach was not only to overspend the budget, but also to promise future considerations that assumed even greater resources would materialize down the road. After all, the benefactor, unlike Johns Hopkins, was still alive and offered some basis for such faith, but Harper was acting as if the money were already committed. Gates, fearing a train wreck, went to Rockefeller seeking additional funds. Although this was not the business model that John D. Rockefeller had envisaged for his charity, he again reluctantly increased his support.

Harper's greatest source for talent was Clark University. Founded just a few years earlier, the Worcester, Massachusetts, institution had made an auspicious beginning by focusing on graduate programs in mathematics, psychology, and selected areas of science. A distinguished research faculty was recruited, but financial pressures soon led to dissension between the founder, Jonas Clark, the administration, and the faculty. When the Harper recruiting mission arrived, the faculty was already on the verge of mutiny. Among the plunder from what became known as "Harper's raid" was Albert A. Michelson, who would later win a Nobel Prize in physics.

Looking back at the success of Johns Hopkins and the University of Chicago it is easy merely to credit Gilman and Harper as leaders who realized that the time was ripe for America to advance in the areas of research and graduate education. Such a view vastly underrates the risks they undertook and the wisdom of their choices in making their projects work. Consider the two respective institutions through the microcosm of their start-up mathematics programs and their impact on the history of mathematics in America. To lead the Hopkins department Gilman selected Sylvester, a recognized scholar but aged, unproven teacher with considerable baggage. Under Sylvester, Hopkins established graduate mathematics education in the United States, instantly surpassing the fledgling programs at venerable institutions such as Yale and Harvard. It would be Chicago, however, that demonstrated that world-class mathematicians could be cultivated in America. The principle figure in the Chicago success was an even more improbable choice than Sylvester.

A century later it is difficult to ascertain Harper's expectations as he approached Eliakim Hastings (E. H.) Moore in 1891 about leading the Chicago mathematics venture. Moore had just turned 29 and was in the second year of an assistant professorship at Northwestern University. For his age, Moore's vita was as impressive as could be expected of an American mathematician. Six years earlier he received his PhD under Hubert Newton at Yale. To his credit, Moore then went to Berlin for a year of postdoctoral seasoning with the German masters. Following a brief stint of high school teaching, Moore returned to Yale, where he continued his research as a tutor in mathematics. While at Yale Moore came into contact with Harper. During those two years, Moore must have made quite an impression. Moore left for a position at Northwestern, but Harper remembered the young man's commitment to scholarship.[3]

When the time came for Harper to select his first mathematician, it was Moore whom he chose to interview. With four papers published since his PhD, Moore deserved better than his current assistant professorship in a department not yet oriented toward research. In fact he was about to be promoted to associate professor. Harper's job description, however, called for a full professor who would prove world-class theorems and supervise the most promising doctoral students. Northwestern did not even have a PhD program. Moore's youth was presumably not too much of a problem for Harper, who himself was only five years older. Moreover, Harper was strapped for funds and Moore would not command the high salary demanded by an already established professor from a prestigious institution. Still, to entrust an inexperienced person with such a vital mission, Harper was placing enormous faith in his personal judgment and whatever assessments he had received from his former colleagues at Yale.

Moore responded favorably to Harper's overtures, but attempted to negotiate for a higher salary, proposing $4,000.[4] Harper held firm at $3,500. This figure was agreed to early in 1892, as Moore became a full professor and acting head professor of the University of Chicago mathematics department. His first task was to organize the new department and obtain personnel in time for the opening of classes in the fall. Moore hit the ground running. He proposed a teaching roster to consist of himself, an associate professor, an assistant professor, and two lower-level tutors such as the position he had held at Yale. Harper scaled back the request to delete one of the assistant/associate level professors.

In recruiting faculty for his department, Moore immediately focused on (Felix) "Klein's men."[5] Over the next few months he raised the names of Henry Seely White, Oskar Bolza, William Osgood, and Maxime Bôcher (the former two at Clark and the latter two at Harvard). In the end Harper found the funds to hire Bolza as an associate professor and Heinrich Maschke as an assistant professor. Bolza became part of the Clark exodus while Maschke left a nonacademic job as an electrician. Bolza and Maschke were close friends who had studied together under Klein at Göttingen. It remained to be seen whether Moore could lead the two older German mathematicians in forming an effective department. The results were astounding.

Moore's research blossomed and he quickly emerged as one of the most prominent mathematicians in the country. Moore kept abreast of all the latest developments and shifted his work through phases in geometry, algebra, and analysis. An indication of his international standing was the conferral of an honorary PhD from Göttingen in 1899. In 1901 he became president of the American Mathematical Society. Two years later he was ranked first in achievement among mathematicians in *American Men of Science*.[6]

Mathematical research thrived throughout the Moore-led Chicago department. Bolza made important advances in the calculus of variations while Maschke obtained fundamental results in algebra. Together the trio offered a cutting-edge array of courses and seminars that rivaled the training available in Göttingen. It was precisely the intellectual environment that Harper envisaged for the new university. By the turn of the century Chicago was training more mathematics PhDs than any university in America.[7]

What was most remarkable was the quality of the educational package that emerged. During the first decade and one half of the mathematics department's existence, a staggering number of five future presidents of the American Mathematical Society received PhDs from the University of Chicago. Moore's students included Leonard Dickson, Oswald Veblen, and George D. Birkhoff, the greatest American mathematicians of the rising generation. These three scholars would attain international prominence from positions in their respective departments at Chicago, Princeton, and Harvard.

The University of Chicago remained the undisputed leader of American mathematics until the death of Maschke in 1908 and return of Bolza to Germany shortly afterward. At this point Chicago lost some of its momentum while Princeton and Harvard advanced with Veblen and Birkhoff, among others. Nonetheless, Moore and Harper, just as Sylvester and Gilman before them, had succeeded in ratcheting American mathematics to a new plateau. More generally Harper, who died in 1906, had established an innovative university that assumed its place in the front rank of higher education.

For the benefactor the success of the University of Chicago was bittersweet. After years of agonizing over whether to fund the project, Rockefeller was gratified by the outcome. This satisfaction was annually eroded by the discovery that, despite repeated admonitions, Harper had overspent the budget. Not only was Rockefeller outraged by this unsound business practice, but there was another aspect to the conflict. Rockefeller's conception for charity was for donations to arise out of his own initiative, rather than as bailouts in response to Harper's personal pleas. The University of Chicago was Rockefeller's first large scale philanthropy ($35 million through 1910) and the experience would shape his future ventures.

Rockefeller had declined to have the university named in his honor, or even to serve as a trustee. His model was to rely on competent managers, assuming that they would be sensitive to his wishes. From his dealings with Harper, Rockefeller realized that he needed a mechanism of exercising more control, while continuing to maintain his separation. It was during the Chicago enterprise that Rockefeller began to experiment with employing Fred Gates as his voice in philanthropic affairs.

Gates, a second-generation Baptist minister, had a resumé that appealed to Rockefeller. He had already advised George Pillsbury on charitable donations. With Pillsbury, Gates engineered the notion of a challenge grant. Gates' combination of business savvy and of sensitivity to spiritual values separated him from the crass charity solicitors who hounded Rockefeller on a daily basis. Gates had already left the pulpit to pursue fundraising for Baptist causes. Shortly after Harper accepted the Chicago presidency, Rockefeller appealed to Gates to become his personal advisor on charitable donations. Gates' task was to screen the interminable pitches and sort out those that were worthy of Rockefeller's attention. From an office in New York Gates began to investigate the appeals. As Rockefeller gained faith in Gates' judgment, the scope of the position broadened. After years of battling Harper's deficit financing, Gates was installed on the University's Board of Trustees and took steps to rein in the spending.[8]

As Rockefeller's filter, Gates was in a position of enormous influence. He had wide discretion and a mandate to recommend the allocation of vast sums of money. It was up to him to nominate the projects most in line with Rockefeller's priorities of improving society "without weakening the moral backbone of the beneficiary."[9] To remain alert to promising new opportunities, Gates read voraciously. One text would serve as the catalyst in first linking the Rockefeller and Flexner families.

It was not the genre of book that made the reading list of entrepreneurs. William Osler's *Principles and Practice of Medicine* was published in 1892 and is considered the first textbook of modern medicine. Its author, one of the

original Johns Hopkins medical faculty, would become the best-known physician of the early twentieth century. Gates devoured the thousand pages during
a summer vacation in 1897. Rather than learning how to practice medicine
Gates came away with a perspective on the state of medical science that has
been described as follows:

> Gates was appalled by the backward state of medicine unintentionally dis-
> closed by Osler's book: while the author delineated the causes of many dis-
> eases, he seldom identified the responsible germs and presented cures for
> only four or five diseases. How could one respect medicine that was so
> strong on anecdote and description but so weak on diagnosis and treatment?
> Gates had a sudden, vivid sense of what could be done by a medical-research
> institution devoted to infectious diseases. His timing was faultless, for major
> strides were being made in bacteriology. For the first time, specific microor-
> ganisms were being isolated as the causes of disease, removing medicine
> forever from the patent-medicine vendors such as [Rockefeller's father].[10]

Diseases that are routinely cured today could then be fatal in the prime
of life. Gates was especially sensitive to the limitations of medical science,
having lost his first wife after just 16 months of marriage. He wrote a memo to
Rockefeller proposing the establishment of a medical research institute. While
such an idea may seem unimaginative today, there was then no precedent in
the United States. In fact Rockefeller, whose name would become identified
with the sponsorship of medical research, gave no immediate reaction to the
proposal.

Rockefeller's reticence was not surprising. He relied on a homeopathic
physician for his own care. The principle of homeopathic medicine was to
treat disease by prescribing small doses of a drug that, when administered
in larger amounts to a healthy person, was known to produce symptoms of
the same disease (for example, quinine for malaria). The competing school
of allopathy advocated employing drugs that induced different effects. It
was amidst this division of American medical schools into homeopathic and
the more traditional allopathic camps that the Johns Hopkins faculty was attempting to pioneer the sciences of bacteriology and immunology. Gates immediately signed on with the Johns Hopkins crowd and faced the resistance
of both the homeopathic and allopathic wings that dominated the medical
establishment.

It would take a few years for Rockefeller to endorse the medical research
institute concept. During this period Gates gained a valuable ally. Rockefeller's only son, John D. Rockefeller, Jr., graduated from college and was

apprenticed to Gates. While learning to deal with the wealth that he was destined to inherit, Junior became an enthusiastic advocate for the medical institute and took advantage of his unique access to lobby his father for the cause. To bolster their case Gates and Junior engaged consultants who explored the operational aspects. Could suitable personnel be recruited? Where would the institute be located and would it be associated with a university? (Any prospect of affiliation with the University of Chicago had evaporated when Harper, to Gates' consternation, merged with the allopathic Rush Medical School.)

In 1901 a Rockefeller family tragedy brought home the need for advances in medical science. John Rockefeller McCormick, Senior's first grandchild, died of scarlet fever. A few months later Senior approved Gates' idea in principle, stipulating a small-scale beginning. The lesson of the University of Chicago had introduced extra caution. Instead of an initial endowment, an annual stipend of $20,000 was authorized for a ten-year period. A panel, consisting entirely of doctors, was constituted with the authority to move forward and the understanding that greater resources might be available in the future. Serving as chair of this group was the most influential American figure in medical research, William Welch of Johns Hopkins.

The early progress was measured. On the personnel front there was an overture to Harvard pathologist Theobald Smith. The first expenditures supported research grants for meritorious investigations in North American laboratories. Establishment of a physical plant was deferred. This was the type of conservative approach preferred by Rockefeller. The organizational plan was shaping up for the Rockefeller Institute for Medical Research. This time Senior permitted his name to be honored, but again he held no formal position. Welch's Board established policy and programmed the allowance provided by Senior. Junior and Gates were the liaisons between the experts and the benefactor.[11]

When Smith elected to remain at Harvard, Welch proposed his former student Simon Flexner for the position of director. It was understood that a more concrete plan would be required to attract Flexner from his secure position at the University of Pennsylvania. Together, Flexner and the Board presented a long-term proposal that included a laboratory, animal facility, and research hospital, along with adequately compensated staff. Senior accepted Junior's recommendation of a five-fold increase to a $1 million commitment over ten years, but without endowment. After Flexner came on board, a site was purchased in 1903 on the east side of Manhattan. The Institute would be an independent operation with neither a university affiliation nor a teaching function.

The timing could not have been more propitious. Science had just reached the point at which it was able to uncover linkages between microorganisms, diseases, and cures. Flexner was at an intermediate point on his ascendance toward a brilliant career. Moreover, with the assistance of Welch, he was in a position to identify the most able scholars. The Rockefeller Institute would begin with personnel on the cutting edge of the discoveries that were finally converting the practice of medicine into a science. When Flexner made a major breakthrough in the treatment of meningitis, it became clear to Senior that the probationary period was over and it was time to establish an endowment.

Along with the endowment it was necessary to set up a system of accountability. Typically, universities were overseen by trustees selected from the business community. This was the case at both Johns Hopkins and the University of Chicago. The Rockefeller Institute took the revolutionary step of converting Welch's advisory group into a board that would control the scientific program. Ultimate authority was vested in a board of trustees that included two representatives from the scientists. Finally, the all important budget was regulated by a committee consisting of three scientists and two trustees. The lay-members of the budget committee became Junior and Gates, meaning that the previous working arrangement was essentially codified. The Rockefellers had broken new ground in permitting scientists a major role in institutional authority.

Simon cast a wide net in recruiting expertise for the Rockefeller Institute. The early group was an international assortment of pathologists, bacteriologists, biochemists, physiologists, and pharmacologists. In 1906 the French-born Alexis Carrel joined the staff to pursue his innovative study of surgical techniques and transplants. When Carrel received the Nobel Prize in 1912, the first time that the award recognized medical research performed in America, it highlighted the extraordinary initial decade of the Rockefeller Institute's existence. Prestigious awards to other staff members would follow as Senior continued to enhance his namesake with donations that would total $61 million. The legacy of the Rockefeller Institute extends beyond its many ground-breaking discoveries. Today the concept of endowing medical research is one of the first considerations for any philanthropist. Winston Churchill commented on Senior's achievement:

> When history passes its final verdict on John D. Rockefeller, it may well
> be that his endowment of research will be recognized as a milestone in the

progress of the race. For the first time, science was given its head; longer term experiment on a large scale has been made practicable, and those who undertake it are freed from the shadow of financial disaster. Science today owes as much to the rich men of generosity and discernment as the art of the Renaissance owes to the patronage of Popes and Princes. Of these rich men, John D. Rockefeller is the supreme type.[12]

Paralleling the development of the Rockefeller Institute for Medical Research was a program motivated by the Rockefeller family's longstanding commitment to the education of southern African-Americans. The General Education Board was begun in 1902 with a ten-year stipend from Senior. Gates chaired the group of supervising trustees that included Daniel Coit Gilman, Junior, and a variety of business leaders. The full-time administrative director was another former Baptist minister, Wallace Buttrick. Under its 1903 Congressional incorporating act, the mission of the General Education Board was broadened to "the promotion of education within the United States without distinction of race, sex, or creed." The first task was to focus on both the white and black races in the southern region. While the intent was admirable, the fact that these were viewed as separate problems with different goals provides some clue as to the patronizing approach and accommodation of racism that would be seen in the Board's actions.[13]

Over time, the General Education Board expanded its initiatives to span the broad range suggested by its name. Just as with the Rockefeller Institute, Senior provided an ample endowment once he was convinced that the Board was properly established. Another parallel was the Rockefeller-Flexner connection that developed when Gates selected Abraham Flexner to become Buttrick's lieutenant. Gates, always on the lookout for an ambitious project that might appeal to Senior, was intrigued by Abe's controversial exposé of American medical education that was published in 1910.

Two years earlier Flexner had returned from his study abroad. He had just completed a short book setting forth his opinions on American college education. At this point Abe was still groping for a career direction. Suddenly he was approached by the other major philanthropic source, the Carnegie Foundation, about conducting a study of medical schools. With no previous training or experience in the practice of medicine, Abe first questioned whether he had been confused with his brother Simon. Assured that the opportunity was intended for him, Abe took up the task and developed a methodology that began with an axiom: the visions of Gilman and Welch at Johns Hopkins were the standard to be emulated. He would make a site visit to each of the

155 medical schools in the United States and Canada, assessing the entrance requirements, facilities, finances, and staff of the program.

Abe's study revealed that Johns Hopkins was the only medical school in the country where an undergraduate degree was a prerequisite for admission. On his visits Flexner discovered schools that purportedly required a high school education, but failed even to enforce that standard. Laboratory facilities and equipment frequently turned out to be less than advertised. Flexner was convinced that such schools were diploma mills functioning to supplement the income of local doctors. These findings were published in his scathing report, accompanied by the conclusion that there was an overproduction of physicians, an appalling number of whom were ill trained. He recommended the closing of most American medical schools, naming names. In one example of the candor in the report, Flexner wrote, "The city of Chicago is in respect to medical education the plague spot of the country." Two Georgia schools were described "as utterly incapable of training doctors" and another as "hopeless."[14]

There was widespread fallout from Abraham Flexner's report. Many medical schools were shut down and others were improved. With his sequel on European medical programs, Flexner became acknowledged as an international authority. Out of his discussions and investigations Flexner began to refine his own view of medical education. He identified the two ingredients essential for a successful program. These were the laboratory infrastructure of a quality university and the clinical environment of a first-class hospital. It is difficult today to appreciate the profundity of this conclusion. In the early twentieth century such infrastructure was rarely associated with medical education.

One fascinated reader of the Flexner report was Fred Gates. Gates saw an opportunity for the General Education Board to upgrade medical education in America. When he asked Flexner to recommend a first step that could be achieved with an infusion of Rockefeller money, Gates was delighted by the answer. Flexner urged that the funds be placed in Welch's hands to enhance the program at Hopkins. Both men had complete confidence in Welch. Moreover the Rockefeller philanthropic philosophy was to "make the peaks higher."[15] If the best were enabled to become better, then competitive forces would compel other medical programs to follow suit.

Thus began Flexner's long association with the General Education Board. In 1911 he was dispatched to Baltimore and conferred with Welch. Out of these discussions Flexner became an advocate of "the full-time approach" to medical education. Welch was salaried by Hopkins to teach and do research. This was the case for all of his colleagues in the laboratory subjects including physiology, pathology and anatomy. The clinical faculty such as surgeons and gynecologists, however, received a part-time teaching stipend and were expected to obtain the majority of their income from private practice. Often they had difficulty dividing their time between the competing needs of patients, students, and research. The full-time approach called for the entire medical faculty to consist of regular salaried professors whose priorities were research and teaching. Any fees collected from patients would then go to the university. This radical change involved the redirection of a great deal of money. It required the Medical School to compensate doctors up-front for the lucrative fees they were forgoing from private practice. Meanwhile the clinicians had less control over their incomes. With Flexner as the go-between, the Johns Hopkins Medical School adopted the full-time approach and the General Education Board subsidized the additional cost.

In 1913 Flexner was appointed to the General Education Board. Over the next several years Flexner selected programs to receive assistance in converting to the full-time approach. Often this designation was accompanied by upgrades in facilities. The goal was to establish regional "peaks" with solid clinical and laboratory resources that would spur a nationwide improvement in medical education. The program was enormously successful.

During the early 1920s Senior provided $50 million for the medical initiative. It was his intention that this money be supplemented by local funds from each community. Flexner made it work. The efforts to secure matching funds connected him with some of the wealthiest families in the country. There was Julius Rosenwald (Sears) for the University of Chicago, George Eastman (Kodak) for the University of Rochester, and J. P. Morgan for the Cornell University Medical School, located in New York city. Combining a smooth, self-assured pitch with respectful deference, Flexner made a strong impression. He succeeded in securing partnerships for the General Education Board with each of these leading entrepreneurs.

While Flexner's charms worked wonders in executive dining rooms and mansion parlors, his dogmatism could wear thin on coworkers and those who were denied funding. When Harvard refused to subscribe to the full-time approach, Flexner embargoed Rockefeller dollars from its medical program. Harvard President A. Lawrence Lowell and influential medical faculty were indignant that a layperson would presume to order a restructuring of their institution. Flexner held firm, leading to a bitter enmity with Lowell.[16]

Fred Gates was the colleague who was responsible for Flexner's appointment to the General Education Board. Gates' relationship of mutual admiration with Simon predated his introduction to Abe. When Abe proposed that the General Education Board fund an enhancement of the University of Iowa Medical School, the retired Gates was still participating in the Board's activities. Gates opposed support to Iowa on the grounds that it was bad policy to support a state university. To Flexner the necessity of establishing a "western" peak left no alternative. The argument was heated. Flexner won the vote and was undoubtedly correct, but Gates was so embittered after the dispute that he withdrew from further Board activities and distanced himself from the younger Flexner brother.[17]

Few people were indifferent toward Flexner, whose influence over the Rockefeller millions gave him enormous power. His opinions went beyond medical training and were harshly critical of a variety of practices in higher education. Given Flexner's limited scholarly background it is easy to understand that university presidents and members of the medical establishment frequently took offense at his outspoken arrogance. Flexner accepted such controversy as the cross that he bore as a reformer. Indeed he is still recognized as a key figure in the history of medical training.

By the mid-1920s the full-time approach had fallen out of fashion. Flexner bristled at retrenchment in medical programs he had championed a few years earlier. He stubbornly attempted to defend his ground, but the Rockefeller Foundation, too, had changed over his 15 years with the organization. When Wallace Buttrick retired in 1923, Flexner lost a dear friend and advocate. Flexner's unhappiness was compounded when he was passed over to succeed Buttrick. An outsider, Wickliffe Rose, became president of the General Education Board. Rose had his own agenda. Part of Rose's program was the establishment of the International Education Board to advance the development of science, including mathematics, on an international basis. Flexner's causes were moved to the back burner.[18]

Under Rose and the new Rockefeller Foundation generation, Flexner's influence and authority diminished. When a Rockefeller reorganization was implemented in 1928, nobody wanted the 62-year-old curmudgeon on their team. It was easy to make the case against such a controversial figure. The Rockefeller enterprises received enough hostility from their monopolistic business practices. They did not need a lightning rod at their charitable offices. In 1928 a front page article appeared in the *New York Times* under the headline "DR. FLEXNER TO QUIT ROCKEFELLER BOARD; ACTION UNEXPLAINED." Abraham Flexner had been forced to retire.[19]

Louis Bamberger, 1939. (Courtesy of Collections of The Newark Museum Archives.)

CHAPTER THREE

FUNDING DREAMS

The foremost benefactor of modern American medical science was John D. Rockefeller who, for his own medical care, relied on the outdated practice of homeopathy. The wherewithal for the erudite Institute for Advanced Study came from another unlikely source. Louis Bamberger dropped out of school at the age of fourteen and went on to become one of the great self-made business success stories at the turn of the twentieth century. Heirless and late in life, he and his sister Caroline (Carrie) Fuld sought a worthy social legacy for their considerable fortune. The search eventually led to consultations with retired Rockefeller Board member Abe Flexner. Out of these meetings would come the Institute for Advanced Study.

Prior to their introduction to Flexner, there was nothing in the Bambergers' lives to suggest an interest in the promotion of scholarly research. Louis and Carrie were born in Baltimore, Louis in 1855 and his sister nine years later. The purchase and sale of merchandise were the family enterprises, with the distinction on the maternal and paternal sides being retail and wholesale respectively. It was at his own initiative that Louis quit school and then went to work at menial tasks in his grandfather's store. Within a couple of years he joined his father and brother in a business that dealt in the wholesale of notions. When he had gained a breadth of experience from the bottom up, Louis was ready to strike out on his own. At the age of 28 he left the family business and moved to New York.[1]

There Bamberger obtained work as a buyer for companies located in Washington and California. He was an astute man whose ambition was cloaked by his quiet demeanor, diminutive physique, and enormous self-control. Given his mercantile background, Bamberger was tempted to launch his own enterprise, but he was well aware of the risks involved in retail trade. He patiently deferred his aspirations while adding to his experience and accumulating the capital he knew would be necessary for a successful venture.

After several years in New York Bamberger began positioning himself to start
his own store. By his own analysis the keys to success were location, capital,
and securing trustworthy associates.

After considerable study of the location problem, Bamberger honed in
on Newark. For months he surveyed the commercial district seeking the perfect site to accommodate his limited resources. Thirty years later Bamberger
recalled his efforts:

> Day after day and night after night I walked the streets, watching the peo-
> ple. I counted the people passing. Broad Street then as now was the main
> shopping district. But I could find no location on Broad Street that met my
> needs. Then I tried Market Street. It was not an attractive street; there were
> many saloons, and the dry-goods shops were mainly of the type that showed
> their goods on the sidewalk. But one block from Broad Street there was
> what was, for those times, a fine building. I thought the people could be
> induced to walk one block if they could get something worth walking for.[2]

It remained to assemble his personnel team, gather additional capital, and
wait for a vacancy. After some time a store space became available, but the
other elements were not yet in place. Bamberger was forced to stand by as
another firm opened the dry-goods business he had been planning. It was a
crushing blow to come so close to realizing his dream, only to have it seemingly put out of reach. With characteristic patience Bamberger continued
his preparations. Four months later the new store on Market Street went
bankrupt. Rather than take this as evidence that the location was unsound,
Bamberger moved quickly to secure the space and stock of the bankrupt firm.
He also sent for the two men whom he had lined up as partners, his brother-
in-law Louis Frank and Felix Fuld (who would later also become his brother-
in-law).

Frank had married Carrie ten years earlier. The couple settled in Philadelphia where Frank's business thrived and then failed. Fuld, who had immi-
grated from Germany at the age of 14, was also keen for a fresh opportunity.
He instantly gave up his job as a traveling salesman and joined Bamberger
and Frank in the new venture. Carrie was enlisted to serve temporarily as
a cashier. The first task was to turn over the liquidated merchandise. With
unusually widespread advertising the discounted stock was readily sold. The
shelves were then replenished with new material and Louis Bamberger's business plan came to fruition. Even the Panic of 1893 could not derail the
impressive start of L. Bamberger and Company. Under able management
Market Street was indeed a solid location.

Caroline Bamberger Fuld, 1940. (Courtesy of Collections of the Newark Museum Archives.)

Other merchants in the same building were experiencing less success. When a basement crockery business floundered, L. Bamberger and Company purchased the stock and expanded its operation. This sequence of events was repeated with a shoe store and eventually Bamberger's store (which became known simply as Bamberger's) occupied the entire building. Its scope became that of a department store. To conduct these acquisitions the firm employed the services of a local attorney, John Hardin, who became one of Louis Bamberger's most trusted advisors.

There would be plenty of work for Hardin as Bamberger's experienced steady growth. The original building was enlarged, and then surrounding structures and property were leased and upgraded. The partners fully appreciated that the success of a department store depended on a symbiotic relationship with its community and staff. They sought to build loyalty by balancing a sensitivity to their clientele with deliberate business practices. This was especially evident in the introduction of new services such as a liberal return policy, delivery, and toll-free calling.

While the original objective had been to entice local residents to walk an extra block, Bamberger's growth ultimately made it the leading department store in New Jersey. With accessibility to Manhattan improving, the future competition was across the Hudson River. Long-term plans were drafted for construction of a large, modern building, and Hardin was commissioned to secure titles to the plots of land in the entire block adjacent to the existing structures. It was during this phase, in 1910, that Louis Frank died.

From the beginning of their retail venture Bamberger, Frank, and Fuld comprised a household as well as a business partnership. Together with Carrie they shared lodging commensurate with their rising social standing and wealth. The new department store building opened in 1912 and Carrie married Fuld the following year. Thus for the remainder of their lives Louis Bamberger, Carrie Bamberger Fuld, and Felix Fuld were intimately bound by the most vital personal and professional connections. Louis Bamberger remained single, and his sister's marriages to his partners were both childless. Without descendants, the devotion of these three concentrated on each other and the business that they had created. It was natural for them to develop paternalistic attitudes toward their customers and employees.

An important ingredient of the firm's success was the determination of Felix and Louis to keep abreast of developments at other stores throughout the country. Employees (called coworkers) were commissioned to visit competitors and return with suggestions for improving company practices. At this point the partners' contrasting personalities functioned to form an effective management team. Fuld was impulsive and frequently eager to pursue new directions. Bamberger was conservative and deliberate. Together they would thrash out new proposals and reach consensus before deciding on implementation. The phenomenal rise of company profits attested to the wisdom of their conclusions.

Their home was located on a 30-acre estate in South Orange, adjacent to Newark. Wealth afforded the Bamberger-Fuld family a variety of luxuries. Their indulgences included widespread travel, the acquisition of paintings, and enjoyment of the opera and theater in New York. Among Louis Bamberger's most prized possessions were a leaf from the first Gutenberg Bible and a collection of signatures that included each signatory of the Declaration of Independence. The family eschewed political activity, but were prominent members of the Jewish community and civic organizations.

Bamberger's became a source of pride and a landmark for the city of Newark. The Bamberger-Fuld family remained mindful that their wealth was derived from the patronage of the local community. Their gratitude was expressed through a variety of charities. The beneficiaries included the Community Chest, Newark Museum, and New Jersey Historical Society. Jewish causes were especially favored, such as Beth Israel Hospital and the retraining of Russian Jewish refugees. However, unlike Rockefeller, the Bamberger-Fuld values always transcended sectarian (as well as gender) boundaries. Not only did they contribute to the Young Men's Hebrew Association and Young Women's Hebrew Association of their own religious faith, but also to the Boy Scouts and Girl Scouts.

These donations were frequently in the hundreds of thousands of dollars. The recipients were carefully selected and reflected the family's gratitude, heritage, and values. It is notable that educational institutions were absent from the list. Culture and art were prominently represented, including the New York Philharmonic Society, a favorite of Carrie Fuld.

As the Fulds and Bamberger aged it was inevitable that some of the household discussions would explore the distribution of their wealth after death. The mounting endowment involved many millions of dollars. Leaving the entire fortune to relatives was ruled out. One priority was to provide for loyal store employees. In those days pension plans were rare and social security did not yet exist. It was agreed that a significant bonus would be paid out to senior employees. Even with this novel stipulation, and bequests to favored relatives and friends, a considerable sum would be available. It was an intriguing problem that had earlier preoccupied Johns Hopkins and John D. Rockefeller, among others. How do you dispose of a fortune so as to produce the best results? One appealing plan was to establish some type of institution on their South Orange estate that would perpetually benefit their constituency, the people of New Jersey.

In January, 1929 Felix Fuld died unexpectedly at the age of 60. This left control of the L. Bamberger and Co. stock in the hands of Fuld's surviving partner and Fuld's wife, Carrie. The prospects for the corporation appeared excellent. In the previous year the store had cleared $3 million in profit from sales of $35 million. Louis Bamberger, approaching his 74th birthday, recognized that he was too old to manage the company on his own. With extraordinarily fortunate timing, he and his sister decided to seek a buyer for their store. The front page of the June 30, 1929, *New York Times* carried a story under the headline "Macy's Acquires L. Bamberger and Co." It would take some time to implement the details that involved turning all the L. Bamberger and Co. stock over to Macy's. In return the sellers would receive a combination of Macy's stock and cash. Late in August the New York Stock Exchange approved Macy's application to list the two blocks of its stock that would be required to consummate the transaction. The Bamberger-Fuld family would then own a $15,000,000 interest in Macy's and receive the proceeds from the sale of stock marketed at a value of $11,000,000.[3]

The timing here is important. The sale carried over into September. This was the month when the stock market in general, and Macy's, in particular, peaked following protracted advances. Both began to decline in October, with the famous crash occurring on October 29. Not only did Bamberger and Fuld receive their cash at the most opportune time, but with any slight delay, their deal would have been jeopardized by the precipitous evaporation of capital. While they were left with a substantial holding of depreciating Macy's stock, they approached the Great Depression having just cashed in for $11,000,000.

Louis Bamberger retired on September 15. The following day he distributed stipends totaling $1 million to employees with 15 years of service. This extraordinary act of generosity attracted a great deal of attention, and Bamberger was quickly targeted by a variety of needy causes and individuals. There was no question that Bamberger and Mrs. Fuld would jointly leave their fortunes to some project aimed at benefiting the people of New Jersey. Before his death Felix Fuld had advocated establishment of a dental school, but Bamberger and Mrs. Fuld preferred the concept of a new medical school. They intended to provide preference in admission for individuals of the Jewish faith, subjects of discrimination in higher education, but the institution would be open to all, regardless of race, religion, or gender.[4]

In philanthropy, as in business, Bamberger knew that an appealing idea required careful study. Before committing his fortune to the launch of a medical school, Bamberger needed assurance that the plan was sound. Moreover, he wanted to separate himself from the inquiry so that he would have complete freedom to act in a different direction. To investigate the feasibility of opening a medical school in Newark, Bamberger turned to two trusted associates. Both were New York professionals, his long-time accountant Samuel Leidesdorf and his more recent financial counselor Herbert Maass.

Leidesdorf and Maass utilized their connections to begin a series of consultations through the New York medical network. They presented themselves as the representatives of unnamed benefactors who were contemplating the endowment of a medical school in Newark. Amidst these discussions and referrals it was inevitable that the name of Abraham Flexner would arise. Not only had Flexner written the most influential book in the history of American medical education, but over the past decade he had orchestrated the massive Rockefeller philanthropy initiative in this direction.

When Leidesdorf and Maass reached him at the end of 1929, Flexner was operating out of an office in the Rockefeller Institute for Medical Research. He had kept busy following his separation from the General Education Board. One and a half years earlier Flexner had spent an extended residence at Oxford University in conjunction with his delivery of the prestigious Rhodes Trust Memorial Lectures. Now he was expanding the talks into a book manuscript, animating his views on the state of higher education.

Flexner's credentials in medical education and philanthropic administration established him as the perfect resource for a vetting of the Bamberger proposal. Leidesdorf and Maass made their pitch and solicited a reaction. It was just the situation where Flexner's charms were overwhelming. Picture his self-assured analysis enlivened with first person anecdotes costarring the Rockefellers, Fred Gates, William Welch, and his brother Simon. Flexner had a gift for smoothly inserting medical jargon and dropping names in a manner that impressed laymen, but concealed his own superficial understanding. Leidesdorf and Maass must have felt privileged to receive a seemingly objective appraisal from the authority who had set the course for American medical education. What they did not know was that the Newark initiative was in conflict with a regional plan that Flexner himself had concocted years earlier.

In his autobiography Flexner proudly recounts his efforts to proselytize the Johns Hopkins Medical School model. The midwest was provided for by an infusion of funds and reorganizations at Washington University and the University of Chicago. It was clear that another regional peak was needed in New York City, but neither Columbia nor Cornell were responsive to Flexner's urging to modernize their medical schools. Out of this frustration he decided "that the situation might be taken in the flank."[5] Flexner successfully solicited George Eastman to partner with Rockefeller in establishing a medical school at the University of Rochester. Seeing the rise of a superior program cross state, the New York City schools fell into line.

There was no Newark node on the map of Flexner's master plan for American medical education. He patiently explained to Leidesdorf and Maass that a modern medical school could not be created in a vacuum. The laboratory and clinical demands required a first-class university and hospital on the premises. Newark simply did not have the requisite infrastructure, and there was no need for a new medical school in the region. As to the notion of admission preference for Jews, Flexner asserted that religious discrimination was not a problem among medical programs.

Flexner was a master of persuasion and he easily disabused his visitors of any consideration of Newark as a suitable site for a medical school. This, however, was not the end of the consultation. Flexner had made his career out of conceiving grand programs and then coaxing millionaires to back them. Now he was within reach of a benefactor seeking a program. This was not an opportunity to be missed. Flexner asked Maass and Leidesdorf if they "had ever dreamed a dream."[6]

Since his college days Flexner's intellectual hero had been Daniel Coit Gilman. While Gilman had created a great university at Johns Hopkins, his original plan had been diluted by public pressure for undergraduate education. Now the university was moving away from its original ideals, and no other institution had filled the void. For years Flexner had been convinced that a true Gilman type university was needed in America, but he had never succeeded in launching such a venture. Working on his manuscript of the Oxford lectures had brought fresh reflection on the problem. Perhaps the mystery philanthropist could be induced to invest in Flexner's dream.

Flexner made his case with an historical perspective on advanced scholarship. German universities led the way through the nineteenth century, but were irreparably harmed by World War I. Gilman had made an impressive start at Hopkins, and admirable graduate programs existed at some American universities. However, undergraduate and graduate education were incompatible. Each diverted the other from accomplishing its mission, contributing to the misshapen priorities that were afflicting universities. In particular, remedial classes, athletic programs, and amenities directed at immature, ill-prepared undergraduates degraded the intellectual atmosphere that was necessary to incubate serious research. What was overdue in America Flexner coined as a "modern university," dedicated exclusively to genuine research and graduate education. All of his philosophy of higher education would be detailed, especially the criticism, in Flexner's book entitled *Universities: American, English, German.*

Leidesdorf and Maass left with a portion of the manuscript and the ball in their court. Bamberger had charged them to determine whether a Newark medical school was feasible. They would report that it was not, but should they transmit Flexner's proposal for an alternative use of the fortune? After all, the only basis was the digression in their discussion with Flexner. That Leidesdorf and Maass did report the entire conversation is a tribute to the forcefulness of the impression that Flexner had made. The insular Bamberger responded by proposing that they all discuss the idea with Flexner and Mrs. Fuld at a dinner meeting.

Bamberger was a cautious man who was making one of the most important decisions of his life. The dinner took place in New York at the Hotel Madison where Bamberger and Fuld shared an apartment during opera season. Flexner was back in his familiar role of pitching proposals to the wealthy. Bamberger was as comfortable a listener as Flexner was a talker. At the end of the evening Bamberger proposed the discussion be continued at weekly luncheons.

The concept sketched by Flexner in January of 1930 was pure Gilman. It was to be a graduate institution in which the only degree to be granted was the PhD. The centerpiece was a well-compensated faculty of the highest caliber with complete freedom to pursue their research while training a small number of students. Under these ideal conditions the faculty would produce groundbreaking advances while their students would go on to positions of leadership in American universities.

Despite the curious casting of a high school dropout being solicited to enable an institution of unprecedented scholarship, Bamberger liked what he heard. It was an original concept for a public benefaction that would operate in perpetuity. It could only be realized with the commitment of a person with his means. While Bamberger was too humble to bestow his name on the institution, he wanted the pleasure of observing its good work at close range. In his mind it was a project ideally suited for his estate, or in its vicinity. There, both Newark and New Jersey would share in the status it accrued. Another vital asset was Flexner himself. This man had a clear vision of the concept and he exuded the *savoir faire* to procure the personnel. Moreover he held the conservative Gilman philosophy that emphasized "men over buildings" and "good work in a limited field."

It was Bamberger's practice to proceed in a deliberate manner. This venture was to be the culmination of his life's work and it was worthy of the time and patience necessary to see it done correctly. He was, however, 74 years old and well aware of his mortality, having recently lost Felix Fuld at a younger age. Bamberger and Mrs. Fuld were about to depart for a winter vacation in Arizona, and it seemed prudent to provide for an unfortunate contingency. Flexner was enlisted to draft a memorandum that embodied the elements of his graduate university. The document would serve as codicil to the Bamberger and Fuld wills. What follows is Flexner's January 20, 1930, draft, providing insight into his own understanding at that stage of development.

MEMORANDUM

It is our purpose to devote our entire residual estate to the endowment of
an institution of higher learning situated in or near the City of Newark
and called after the State of New Jersey in grateful recognition of the opportunities which we have enjoyed in that community. We are persuaded
that there is now little or no lack and that there will in the future be still
less lack of schools and colleges for the training of young men and women.
Neither now nor in the future is there likely to be an overabundance of
opportunities for men and women competent to advance learning in all serious fields of human interest and endeavor and to train younger men and
women who may follow in their footsteps. It is our desire therefore that
the proposed university shall contain no undergraduate department, that as
long as present conditions continue, it shall bestow only the PhD. degree or
professional degrees of equal value, and that its standards of admission and
its methods of work be adapted to these ends and these only. As conditions
in the realm of advanced instruction and research improve, it is our desire
that the trustees of this institution advance the ideals of the institution so
that it may at all times be distinguished for quality and at no time be influenced by consideration of numbers. It is our express and inflexible desire
that the appointments of staff and faculty of this institution and in the admission of workers and students no account be taken directly or indirectly
of religion or sex. In the spirit characteristic of America at its noblest, we desire that this fund be administered with sole respect to the objects for which
it is set up and with no respect whatsoever to accidents of creed, origin, or
sex. It is our belief that the sum which we shall ultimately provide will be
adequate to start and to maintain at the highest possible intellectual level an
institution devoted to the central cultural and scientific disciplines. It is no
part of our immediate intention to institute professional schools. It is our
wish that our trustees should not countenance development in that or any
other direction unless funds are assured which permit the undertaking of
additional responsibilities at the same high level at which the enterprise has
been started. It will probably develop that most candidates for the doctor's
degree will have received a collegiate degree or the equivalent thereof, but
it is our wish that the facilities of the institution will be open to any student who can demonstrate fitness to profit in the highest degree by their use
and to no others. It is also our purpose that many of those who enter the
university which we propose to establish will hope to become professors in
other institutions of learning, but we desire to emphasize the fact that the
institution itself is set up not to train teachers, not to produce holders of
degrees, but to advance learning and to train persons competent to participate in that fundamental and most important endeavor. For the execution
of this purpose we temporarily create a committee made up of

In the event of the death of both of us before further steps can be taken, this committee is authorized to constitute itself as the first Board of Trustees by adding to its number ___ members. We commend to their consideration as representing the ideals of scholarship and service to humanity that we have in mind the following:

It is our hope that site, buildings, and equipment can be provided without impairment of the capital sum with which the institution will be endowed. No gifts from outside sources shall ever be accepted conditioned upon the modification of the fundamental aim for which this institution is created. It is our hope that the most cordial and cooperative relations may at all times exist between the trustees and the faculty of the university. To that end we suggest that at least three members of the faculty of the university be chosen ultimately by the faculty itself to become members of the board of trustees, and we further hope that the opportunities of the institution may prove attractive to men of the most distinguished standing because of the freedom and abundance of opportunities which they will enjoy in the prosecution of their own work and in the selection and training of students and in the maintenance of the highest possible standards in science and scholarship.

1. The buildings should be modest, adaptable to their purpose and yet sufficiently attractive to exercise a beneficial influence on the architectural taste of the community.

2. The trustees shall be empowered to establish within reasonable limits such fellowships and scholarships as from time to time may be needed in order to support in whole or part younger men and women whose previous training has been adequate and whose development promises to be significant.[7]

While much of the substance owes to Flexner's discipleship of Gilman, there were several customized aspects. The nondiscriminatory policy dictated by Bamberger and Fuld contrasted sharply with Gilman's feeble actions half a century earlier. The inclusion of faculty among the trustees was promoted by Flexner, whose abstract views of faculty/administration roles were, at this point, egalitarian. It was the issue of location, however, where Flexner elected to patronize his sponsors. In his view Newark was neither a satisfactory site for a medical school nor for the institute that he envisioned. Fully understanding the importance to Bamberger and Fuld of linking the project with their community, Flexner opened the memorandum with a gratuitous declaration for Newark. Upon review, Flexner's draft was truncated, deleting all that succeeded the statement on faculty trustees. A new paragraph was appended, providing flexibility to shift the resources to some other "beneficent public purpose" that operated under a policy of inclusiveness.

Bamberger and Fuld departed for Arizona, where they would have ample opportunity to contemplate their benefaction. They left their opera box to Flexner, who would devote his energy to writing his manuscript on higher education. For the remainder of the winter there was only token correspondence about the graduate university. When Maass returned from his own vacation, he sent a note to Flexner suggesting that they get together for lunch. Flexner replied enthusiastically, proposing to include Leidesdorf. Flexner, who was conducting a restrained campaign for his project, took advantage of the opening to send both men a book that supported his views on university reform. Certainly Flexner was anxious to send the same material to Bamberger and Fuld, but he was waiting for them to make the first overture. Two weeks later a postcard arrived from Arizona, and Flexner responded with the book and a friendly letter.[8]

Meanwhile Flexner was progressing on his manuscript, which he completed and sent to the publisher. He had arranged for multiple copies of the galley proofs to be produced. These were sent to readers at home and abroad, each of whom was requested to fill the margin with their criticisms. In mid-May Flexner planned to sail for Europe, where he would personally retrieve the annotated copies from his overseas critics and discuss their suggestions. After receiving the domestic reaction he would be in a position to make final revisions over the summer.

A few weeks prior to Flexner's departure, Bamberger and Fuld returned from their southwestern vacation. They had given careful consideration to the Flexner concept, as well as to alternative possibilities. Bamberger and Fuld decided to cast their lot with Flexner, under the stipulation that he personally bring the project to fruition. This commitment, in April 1930, was a notable event in American intellectual history. Never before had such resources been devoted, without encumberment, to advance scholarship at the highest level.

It may seem curious that this timing coincides with the beginning of the Great Depression. Not because of any cause and effect, but rather the value system that favored intellectual ideals in the face of pervasive deprivation. Such a judgment, however, is unfair. While indeed the Depression was well under way, its duration and severity were poorly understood at the time. Economic theory of the day held that occasional panics were an inevitable ingredient of the business cycle. For decades these down periods had appeared and then passed (Bamberger began his store during the Panic of 1893). In the spring of 1930 there was little expectation that the recent problems would be any different. Even the stock market had regained some of the ground that it had lost in the crash six months earlier. On May 1 President Hoover declared: "I am convinced we have passed the worst and with continued effort we shall rapidly recover."[9] Such optimism, as absurd as it appears in retrospect, indicates that Bamberger's decision was made in the face of economic conditions that were perceived as adverse, but not extraordinary. His choice, with Mrs. Fuld, of a beneficiary was unquestionably motivated by a sincere desire to effectuate some benefit for society.

The personnel aspect of their move is also worthy of further examination. Despite the fact that Flexner had never held a university position nor published a scholarly paper, he was the only conceivable selection to direct the new university. It was his vision that had inspired the cautious Bamberger to make the huge investment. Bamberger was not going to entrust his legacy to a stranger, and no one among his inner circle was even remotely qualified for the task.

As Flexner moved toward his new career he was 63 years old with a mixed view of his own accomplishments. He looked back with justifiable pride at his vital contributions to American medical education, but Flexner's intellectual values had been instilled at Johns Hopkins. Unlike his brother Simon, Abe had failed to produce the deep research that was the mark of a true scholar. His new role would bring him closer to the actual research endeavor, and possibly provide the satisfaction he lacked.

Through his foundation work Flexner was a well-connected administrator. He had given considerable thought to all aspects of university life, concluding that a great deal was wrong with higher education and in need of reform. In Flexner's forthcoming book, as in his previous review of medical programs, he was set to name names of programs that were astray. The Harvard Graduate School of Business was "pretentious" and "dangerous," and the Columbia University School of Journalism was "placed on a par with university faculties of cookery and clothing."[10] With such scathing commentaries, Flexner was aware that he now had counterparts who would love to watch the chip fall from his shoulder. There would be widespread scrutiny of whether the new institution achieved the lofty threshold that he himself had set. Despite his lack of experience, Flexner was convinced that he could properly administer a graduate university. After all, like Gilman, he could evaluate men.

Flexner and Bamberger had substantial matters to settle in the short period prior to Flexner's departure for Europe. There was a corporation to charter, board members to select, and a mission statement to produce. Even a name was needed and, of course, there was the size of the initial endowment. Perhaps most important was the establishment of a working relationship between the two men. Flexner had already succeeded by being respectful and solicitous. He would continue this posture as he attempted to educate the benefactors and keep them informed of his progress. Recommendations would always be presented as just that, with no pressure such as exerted by Harper on Rockefeller. That said, Flexner would carefully filter the information the founders received, skillfully providing his own spin and monotonously transmitting every compliment that he received.

Few people had more experience than Flexner in discussing multimillion-dollar contributions. The strategy was to obtain as much as possible, while maintaining the good will that would enable future additions. This was especially important with Bamberger and Fuld, who were disposed to leave their residual estates to the project. The initial endowment was settled at $5 million. It was a generous sum that was more than sufficient to begin.

Flexner's compensation was set at $20,000 plus travel expenses and a private secretary. The figure had ramifications for the faculty he was about to recruit. One of the reforms Flexner advocated in *Universities* was the increase of professorial salaries to the level of presidents. This meant that Bamberger and Flexner were contemplating shaking up a market where there was a ceiling of $12,000 at Harvard and Columbia.[11]

For some time Flexner had been groping for a term to describe the type of institution he was advocating. In a 1925 article entitled "A Modern University," Flexner lamented that in the United States the term "university" was ambiguously applied to a spectrum of institutions that included colleges offering only undergraduate degrees, colleges with associated graduate schools and professional programs, and what he characterized as "an educational department store containing a kindergarten at one end and Nobel Prize winners (or as good) at the other." While a new model for higher education was needed, the term "university" was so engrained in educational usage that he concluded it was futile to adopt a new noun. At that time he used "modern university" in the nomenclature for his exclusively graduate concept of higher education.[12]

Several years later, as Flexner enumerated his criticisms in *Universities* and pondered the possibility of a Bamberger-Fuld endowment, he came to believe that there was too much baggage associated to the word and wrote:

> It has, however, become a question whether the term "university" can be saved or is even worth saving. Why should it not continue to be used in

order to indicate the formless and incongruous activities—good, bad, and indifferent—which I have described in this chapter? If indeed "university" is to mean, as Columbia announces, a "public service institution," then the university has become a different thing, a thing which may have its uses, but is assuredly no longer a university. In this event, in order to signify the idea of a real university, a new term may be required—perhaps a school or institute of higher learning—which would automatically shut out the low-grade activities with which institutions of learning have no concern.[13]

Complicating his terminology problem was the need to separate his concept from that of a "research institute" where the teaching component was absent. When Bamberger came on board in April 1930, Flexner toyed with the phrases "Institute of Higher Learning" and "Institute of Advanced Studies." By Bamberger's May 5 formal offer of the Directorship, the preposition had been modified so that the name became "Institute for Higher Learning." It was in Flexner's May 9 acceptance that "Institute for Advanced Study" makes it first appearance.[14]

The evolution of the name and terminology reflected just one of the considerations that arose in crafting a letter of invitation to trustees. The body of this letter was intended to set forth the mission of the institution. Flexner began work on his draft by adapting the language of the codicil. Amidst a variety of cosmetic and syntactical modifications, Flexner engaged in an act of deception. He slyly deleted the reference to Newark as an intended venue, merely stipulating a location "in the State of New Jersey."[15]

About this time Bamberger brought his close friend and advisor John Hardin into the discussions. Hardin, who was then president of the Mutual Benefit Life Insurance Company, was also a trustee of his alma mater, Princeton University. Hardin polished the letter to the trustees as Bamberger weakened the language committing his residual estate. The omission of Newark reached the final draft below that, otherwise, provides a clear statement of the mission as it was understood by both Flexner and Bamberger when the papers for incorporation of the Institute for Advanced Study were signed on May 20, 1930.

Dear Sir:

We are asking you to serve with us as trustees of an institution of higher learning which we propose to endow with a substantial initial sum, to which we expect from time to time hereafter to add amounts which in our belief will provide adequately for the establishment of the proposed enterprise.

There is at present little or no lack of schools and colleges for the training of young men and women for the ordinary baccalaureate degrees. This need will in the future be apparently even more fully supplied than at present. There are also attached to many of our colleges post-graduate schools doing effective work in guiding students in qualifying themselves for post-graduate degrees.

There is never likely to be an over-abundance of opportunities for men and women engaged in the pursuit of advanced learning in the various fields of human knowledge. Particularly, so far as we are aware, there is no institution in the United States where scientists and scholars devote themselves at the same time to serious research and to the training of competent post-graduate students entirely independently of and separated from both the charms and diversions inseparable from an institution the major interest of which is the teaching of undergraduates.

It is our desire, therefore, that the proposed institution shall contain no undergraduate department and that it shall bestow only the Ph. D. degree, or professional degrees of equal value, and that its standards of admission and methods of work shall be upon such a basis and upon that alone.

In so far as students are concerned, it is our hope that the trustees of the institution will advance the ideals upon which it is founded in such manner that quality of work rather than number of students shall be the distinguishing characteristic of enrollment.

It is our hope that the staff of the Institution will consist exclusively of men and women of the highest standing in their respective fields of learning, attracted to this institution through its appeal as an opportunity for the serious pursuit of advanced study and because of the detachment it is hoped to secure from outside distractions.

It is fundamental in our purpose, and our express desire, that in the appointments to the staff and faculty as well as in the admission of workers and students, no account shall be taken directly or indirectly, of race, religion, or sex. We feel strongly that the spirit characteristic of America at its noblest, above all the pursuit of higher learning, cannot admit of any conditions as to personnel other than those designed to promote the objects for which this institution is established, and particularly with no regard whatever to accidents of race, creed, or sex.

In endowing this Institution we recognize that many worthy and capable persons are unable for financial reasons to pursue study or research to the extent justified by their capacities. It is expected, therefore, that the Institute will supply means whereby, through scholarships or fellowships, such workers may be supported during the course of their work or research, to the end that the facilities of the Institution may be available to any man or woman otherwise acceptable possessing the necessary mental and moral equipment.

While the Institution will devote itself to the teaching of qualified advanced students, it is our desire that those who are assembled in the faculty or staff of the Institution may enjoy the most favorable opportunities for continuing research or investigations in their particular field or specialty, and that the utmost liberty of action shall be afforded the said faculty or staff to that end.

It is not part of our immediate plan to create a professional school, and we do not contemplate that the trustees will sanction the development of the Institution in that or any other direction unless separate funds are assured which permit the undertaking of additional responsibilities upon the high level at which the enterprise is started and consistently with the whole spirit of the undertaking.

It will doubtless develop that most of the students admitted to this Institution for the purpose of obtaining a doctor's degree will before entering have received a baccalaureate degree or the equivalent thereof. The facilities of the Institution should, however, in the discretion of the trustees and staff, be open to any acceptable student who may demonstrate his or her qualifications and fitness.

Many of those who enter the Institution will probably qualify themselves for professorships in other institutions of learning, but the Institution itself is established not merely to train teachers or to produce holders of advanced degrees. The primary purpose is the pursuit of advanced learning and exploration in fields of pure sciences and high scholarship to the utmost degree that the facilities of the Institution and the ability of the faculty and students will permit.

It is intended that the proposed institution be known as "THE INSTITUTE FOR ADVANCED STUDY", and, in grateful recognition of the opportunities which we personally have enjoyed in this community, that it be located in the State of New Jersey.

It is our hope that the site, buildings, and equipment can be provided without impairment of the capital sum with which the INSTITUTE FOR ADVANCED STUDY will be endowed.

It is our express wish that gifts from outside sources shall never be accepted conditioned upon any modification of the fundamental aim for which this institution is created.

To the end that the most cordial and co-operative relations may at all times exist between trustees and the faculty of the Institute, it is our further desire that certain members of the faculty shall be chosen to become members of the board of trustees.

This letter is written in order to convey to the trustees the conception which we hope the Institute may realize, but we do not wish it or any part of it to hamper or restrict our trustees in their complete freedom of action in years to come, if their experience with changing social needs and conditions shall appear to require a departure from the details to which we have herein drawn attention.

Faithfully yours,

Louis Bamberger

Carrie B F Fuld[16]

Flexner finally had the opportunity to implement the plan of his (and Gilman's) dreams. He was, however, beginning with a number of enormous advantages over Gilman and Harper. There was no pressure to organize in the face of an approaching deadline for the opening. Bamberger and Flexner were in complete agreement that the programs be developed and faculty recruited at whatever pace proved advantageous. This could mean two, three, or even more years.

Flexner, unlike Gilman and Harper, was in a position to offer salaries that were well above the market rate. Moreover, he had a clear understanding that the student-faculty ratio could be a few to one. Gilman and Harper were managing what was viewed by their community as public trusts. While the Institute for Advanced Study might well open with, say, three faculty and seven students, such numbers would have been unthinkable at Hopkins or Chicago.

This demographic of just a few students, some receiving scholarships, meant that the Institute for Advanced Study would have virtually no income other than from endowment. Budgeting $20,000 chunks for each professor, Flexner was well aware that, in the long run, more money would be necessary. If he could maintain Bamberger's confidence and good will, there was every reason to believe that this funding would be forthcoming. At this point he was unaware that the country was entering the worst economic decade of its history.

From his work with Gates, Flexner undoubtedly knew how Harper's actions at the University of Chicago had antagonized Rockefeller. It was essential to maintain the best relations with Bamberger. The most likely source of contention was the issue of location. Flexner would ultimately need to find a solution that avoided a direct confrontation. He was already plotting to take the issue on the flank.

Another sensitive matter was the selection of trustees. University boards, such as Chicago and Hopkins, were traditionally dominated by nonscholarly professionals who occupied positions of influence in the community. Simon Flexner's Rockefeller Institute had broken new ground by vesting budget control in the hands of experts. Simon was convinced that this enlightened oversight was a critical factor in his administrative success.[17]

Abe had some hopes that the Institute for Advanced Study might be governed under similar circumstances. His slate of nominees for the trustee positions consisted of academic administrators and scholars as well as two ambassadors, a philanthropist, and a corporate president. Bamberger, while willing to dress up the committee with some prominent names, was determined to recruit heavily from his own team. The record provides no evidence of any dissent by Flexner. Several factors, in addition to his deference to Bamberger, may account for Flexner's acquiescence.[18]

On a political level, all of Bamberger's intimates could be expected to follow his lead. Since persuading Bamberger was already a prerequisite for the implementation of policy, the lay-members were unlikely to exert any adverse impact over the final outcome. As for the expert advice that might be available from more knowledgeable candidates, Flexner could just as well solicit outside help from handpicked consultants. Finally, Flexner had considerable confidence in his powers of diplomacy, flattery, and persuasion. Handling the Board should not be a problem, and, for many years, it was not.

The first Board consisted of 15 members that included Bamberger, Mrs. Fuld, Flexner, Maass, Leidesdorf, and Hardin. Two other trustees were executives from Bamberger's store. They were Vice President Edgar Bamberger, a nephew, and President Percy Straus, from the family that controlled Macy's. The remaining seven members, while constituting what might have been a suitable slate for a medical school, were an odd selection for an institution that specifically eschewed the medical field. A majority were physicians or medical researchers.

It was understandable to include Lewis Weed, who as dean of the Johns Hopkins Medical School was a prominent administrator. As well, Alexis Carrel contributed the prestige of a Nobel laureate. Completing the medical wing were Carrel's Rockefeller colleague Florence Sabin and the Baltimore gastroenterologist Julius Friedenwald, a friend and classmate of Flexner from their college days together at Hopkins. Two of the other members were distinguished government figures from well-connected families. There was Herbert Lehman, who was lieutenant governor of New York and a partner

Picture from the first Board of Trustees meeting for the Institute for Advanced Study: from left to right, seated, Alanson Houghton, Caroline Bamberger Fuld, Louis Bamberger, Florence R. Sabin, Abraham Flexner; standing, Edgar S. Bamberger, Herbert H. Maass, Samuel D. Leidesdorf, Lewis H. Weed, John R. Hardin, Percy Seldon Straus, Julius Friedenwald, Frank Aydelotte, Alexis Carrel. (Courtesy of the Archives of the Institute for Advanced Study.)

in the Lehman Brothers firm that brokered the Bamberger's-Macy's transaction. Flexner's friend Alanson Houghton, from the Corning Glass family, had served as ambassador to Germany and to England.

While it remained unclear as to which research disciplines would be represented at the Institute for Advanced Study, one fact was certain. They would be outside the expertise of fourteen of the fifteen board members, including the Director. The one possible exception was Frank Aydelotte. The connection between Flexner and Aydelotte dated back a quarter century to when they met in Louisville. Aydelotte became a Rhodes Scholar who, following his time at Oxford, was an enthusiastic proponent of the British college system. He held faculty positions in the English departments at Indiana University and the Massachusetts Institute of Technology, where he created courses, performed research, and moved toward administration. In 1921 Aydelotte was installed as president of Swarthmore College. Under his leadership Swarthmore gained prestige and became the first American college to offer its students a version of the British honors program.[19]

The Institute Board, though carrying the impressive credentials of Carrel, Houghton, and Weed, was largely a product of the old boy network. Despite the unusual presence of two women trustees, the dominant profile was that of a male who was a longtime associate of Bamberger or Flexner. They were neither going to micromanage nor debate the larger issues. Bamberger himself was there, primarily, to exert his control and observe the unfolding of his philanthropic act. The entire burden of making the Institute work was on Flexner, and he knew it.

In the few weeks preceding his May 14, 1930, departure for Europe, Flexner had secured the funds and put in place the ground work for what he intended to be the capstone of his career. He was now walking in the footsteps of Gilman. There had been little opportunity to really reflect on the details of implementation. Hoping to graft some flesh to his skeletal plan for the Institute, Flexner added a secondary objective to his European consultations. After discussing the *Universities* manuscript, he would seek advice on planning the Institute.

Flexner made the transoceanic voyage and visited his readers in Germany and England. He was satisfied with the feedback on the book. The chapter on British universities was controversial but solid, just what he wanted. The German chapter required some updating to deal with political developments. The consultations on the Institute, however, were more superficial. The concept was embraced, but the reactions were cautious. Writing to Simon from England, Abe summarized the task that lay ahead. "I realise fully that everything depends, as it depended in Baltimore in 1876, on bringing together a group of persons. If I can get them it will succeed—if I can't, it won't."[20]

Flexner returned from Europe early in July, meeting Bamberger and Fuld for lunch. Nobody was in a rush to open the Institute. It was agreed that Flexner would retreat to his summer home in the Canadian woods to complete the book revisions and relax. When this was accomplished he was ready to shift into the role of Director.[21]

Looming ahead was the first trustees' meeting, scheduled for October. At this gathering the tone would be set for the relationship between the Board and Director. It was essential that Flexner make a strong impression and that he establish his authority. With a more scholarly audience he might have raised the fundamental questions of faculty personnel and subject areas. A discussion among the 15 trustees was unlikely to be productive and could cause harm by reaching undesirable conclusions. Flexner elected to defer these issues until his own preferences were better determined. As well, the thorny location problem threatened to provoke the worst conceivable scenario, a dispute with Bamberger.

For the agenda it would be best to stick to back patting and the one pressing item of business, adoption of by-laws. Flexner set to work codifying the structure of the Institute and how he wanted it to function. Daniel Coit Gilman was Flexner's muse, but Wallace Buttrick was his mentor. In directing the General Education Board Buttrick liked to say that "our one policy is to have no policy."[22] Flexner took these words to heart in conceptualizing the governance of the Institute. He saw no reason to adopt a comprehensive set of regulations. Such structures limit flexibility and were unnecessary for serious scholars. The Institute would rely on the wisdom of a faculty and director acting in the best interests of advancing the frontiers of knowledge.

Flexner produced a vanilla draft of the by-laws, establishing committees and defining roles in a general manner. The most striking aspect was a significant voice for the faculty, both in consultation on policy and in holding positions on the Board. Notably absent was any reference to the location of the Institute. As with the trustee letter Flexner was preparing the record to accommodate a later move. Prior to the meeting Flexner reviewed his work with Bamberger and Hardin. Hardin and his son Charles, who was now handling the legal work, produced a revision to address Bamberger's concerns. The faculty role was weakened and a new Article I spoke to location.[23]

The Board made some slight additional modifications, but Flexner's location ploy had failed. The language "at or in the vicinity of Newark" was officially adopted. As to the faculty role, it is not surprising that the department store executives were reluctant to empower the *workers*. The by-laws placed the Director firmly in charge of policy and faculty appointments, without sanction of faculty consultation.[24]

A regular feature of the Board meetings was the presentation of a prepared report by Flexner. At the first meeting he spoke of the grave responsibility of creating a new institution. Two of the themes were slow and small. It was too soon to be specific about fields. The most immediate task was to search throughout the country and Western Europe for the first members of the faculty. "Where and how it begins must depend, in the first instance, on the men and women of genius, of unusual talent, and of high devotion, who may be found willing to be associated with us."[25]

Flexner did elaborate on the circumstances that must be established to attract outstanding scholars to the Institute. It was essential to provide an environment that met both their intellectual and personal needs. For years

Flexner had listened, with sympathy, to professors complaining about their plight of too much teaching, tedious service duties, and inadequate remuneration. At the Institute each professor was to have just two responsibilities: research and direction of a few advanced students. Salaries must be sufficient to ensure a comfortable family life.

The typical student would be a bright, mature, self-motivated individual who had already completed an undergraduate program. The baccalaureate, however, was not essential. If a candidate, lacking a degree, were adjudged as suitably prepared to undertake the rigors of Institute training, then he or she was to be admitted. Postdoctoral students were to be welcome as well. At this stage, however, Flexner expected just a small number of exceptions on either end.

With the organizational Board meeting behind him, Flexner's immediate steps were to set up an office in Manhattan and publish a bulletin on the organization and purpose of the Institute. During this period he celebrated his 64th birthday. Flexner remained vigorous, but was aware that in the approaching months he would have to endure a challenging travel schedule at home and abroad. In his speech to the trustees the phrase "coming near the end of my active career" appeared in an early draft, but was deleted.[26]

The first Institute bulletin consisted primarily of the founders' letter and Flexner's report to the trustees. When Simon read the document, he reacted to his brother's cavalier usage of the term "genius." With the possible exception of Rowland, Simon had seen no such "animals" among the Johns Hopkins faculty, and certainly not in the Medical School. He doubted whether any contemporary name belonged in the genius class.[27] Abe responded with a narrow defense of his choice of words:

> I used the word "Genius" in a Pickwickian sense. Of course it might be restricted to persons like Shakespeare, Sir Isaac Newton and Faraday, but it may also be applied to persons of less fundamental gifts who have a peculiar genius for doing something extremely well. In this lesser sense I should say that Gildersleeve and Sylvester and Dr. Welch had genius, though none of them were geniuses.
>
> You will also notice that I am not pinning my faith even to people who have a certain sort of genius, since in the same connection I mentioned "unusual talent" and "devotion." Whether we shall ever get hold of a genius is an absolute gamble—certainly we cannot count upon it. We ought to have a high proportion of talent and devotion.[28]

George David Birkhoff, 1925. (Courtesy of the American Mathematical Society.)

- CHAPTER FOUR -

DECIDING WHERE TO START

In December 1930 Flexner began to focus his attention on the central issue of faculty selection and recruitment. He often thought about Gilman having secured Sylvester, Rowland, Welch, and Gildersleeve. That was the standard to emulate, and Flexner had in mind several talented individuals for the faculty. The first hires, however, needed to make an even stronger statement. Flexner wanted the Institute to open as a world power in some particular discipline. Moreover, it should be a subject in which no American university were strong. Flexner was remarkably open-minded as to the choice of field. He was willing to set aside his own interests, so long as the Institute demonstrably lifted American scholarship to a higher level.

Before selecting his first area of concentration Flexner would cast a wide net in his consultations. He hoped to obtain some grasp of the fundamental problems of each discipline, as well as an understanding of who was most fruitfully attacking them at home and abroad. His plan was to begin with his own Board and contacts in New York, then travel to the intellectual centers of the United States, and finally sail to Western Europe. All the while, just as Gilman before him, Flexner would promote the Institute.

Among the Board, Aydelotte was likely to possess the broadest view. He and Flexner were old friends who held each other in high esteem. They had originally met in Louisville. There Flexner gave Aydelotte advice on how to learn Greek in preparing his candidacy for a Rhodes Scholarship. When Aydelotte became president of Swarthmore, Flexner was helpful in steering foundation funds to the college. More recently Aydelotte had nominated Flexner to deliver the Rhodes Lectures at Oxford.

Flexner went to Swarthmore, where he had a good discussion with Aydelotte. Upon returning to New York Flexner drafted eleven questions on the initial organization of the Institute. The list began, "With what subjects should the Institute first cope" and "In reference to the subjects selected do you know persons who could undertake to head them." Other questions dealt with students, infrastructure, and salaries. The list was sent to Aydelotte, Sabin, and Weed, the only three trustees with professorial experience. Despite his own medical interests Weed gave the highest priority to history. Aydelotte listed several specific areas of the humanities and social sciences while Sabin singled out subjects in the social sciences and sciences.[1]

Aydelotte and Weed remarked that the sciences had already received preferential treatment in the United States. This view suited Flexner, whose concern was that scientific research must be accompanied by expensive equipment and large laboratories. He believed in "men over buildings" and was determined to place the bulk of his resources into salaries. Mathematical research, however, only required an office, a blackboard, and chalk.

Flexner knew absolutely nothing about mathematics. At an early stage he was seeking out Columbia University faculty for their views on a variety of subjects. David Eugene Smith was happy to provide guidance on mathematics. Smith was an elder statesman of the American Mathematical Society, having served as a vice president and long-term librarian of the organization. Flexner came away from their discussion with the notes that Leonard Dickson of Chicago was the only American "mathematical genius," though the younger G. D. Birkhoff of Harvard "may prove he is able."[2] Smith promised to make further inquiries and remain in touch. For Flexner, the most intriguing revelation may have been that the standing of the American mathematical community was not so high. This impression, confirmed in subsequent consultations, suggested a significant opportunity for the Institute.

Mathematics was by no means the only subject to receive serious consideration, but it had moved to the front of the sciences. At the January 16, 1931, meeting of the trustees Flexner reported that he was progressing, but asked for indulgence. His most revealing remark was that disciplines had developed to the point where no single individual could be expected to cover an entire field, as had each of the original six professors at Hopkins. It might be necessary to begin with greater specialization than the normal classifications such as mathematics or history. Another possibility was to select two or more professors in the same area. Flexner would continue to study the matter and hoped to have specific recommendations at the next meeting, but he gave no clue as to areas or names.[3]

Flexner's later actions indicate that, with the exception of Aydelotte, he had given up hope of receiving any substantive guidance from the trustees. He would keep Bamberger informed of his progress and ask for his approval at key stages. Maass and to a lesser extent Leidesdorf, whom he viewed as allies in persuading Bamberger, received periodic bulletins as well. The remainder of the trustees would learn of actions at Board meetings, essentially as *faits accompli* to rubber stamp.

Late in January Flexner traveled to Washington and Princeton as the consultations moved into their next phase. His correspondence during this period refers to extensive notes that he was making from his interviews. These notes were among neither Flexner's personal papers at the Library of Congress nor the records in the Institute archives. An illuminating, but incomplete, sense of his thinking is derived from his correspondence with family members, Bamberger, and consultants.[4]

In Washington Flexner explored various initiatives in the social sciences. He discussed economics with Supreme Court Justice Louis Brandeis and history with experts at the Library of Congress. Also on the trip Flexner sought out maverick historian Charles Beard. Beard's controversial books had argued that the actions of the Founding Fathers were driven by economic rather than idealistic motives. Fifteen years earlier Beard's principled stands on the United States' involvement in World War I had generated considerable resentment. After resigning in protest from a faculty position at Columbia, Beard was a founder of the New School of Social Research. Although his views on social and political activity were widely regarded as subversive, Beard was one of the most distinguished historians in the United States. Flexner was drawn to his broad command of the social sciences. They established a strong rapport and Flexner began to seek Beard's approval. Such associations were significant, as Flexner's own intellectual insecurity was driving him to create a network of authorities upon whose wisdom he could rely.[5]

This would become especially evident in the consideration of economics. Flexner was impressed by Wall Street financier Paul Warburg. Warburg was a member of an influential German Jewish family. Shortly after immigrating to the United States, Warburg began conceiving the establishment of a central bank to alleviate the irregular panics that seemed ingrained in the American economy. Warburg played a crucial role in the founding of the Federal Reserve Bank, serving as the first vice-governor of the Board. Despite his insider standing, Warburg came to view the implementation of the Fed system as hopelessly flawed by the primacy of political influence over sound banking decision-making. Contributing to his bitterness were the political factors leading to his own separation from the Board in 1918.[6]

Flexner must have been aware that Warburg had foreseen the 1929 stock market crash and the ongoing problems. When they discussed the possible avenues for the Institute, Flexner was struck by Warburg's incisive grasp of the

Charles Beard ca. 1933. (Courtesy of the American Historical Association Archives.)

world economic situation. Out of their conversation Flexner envisioned the possibility of a theoretical economist in residence at the Institute who would, from time to time, venture out into the real world, gaining practical experience from contacts provided by Warburg and Brandeis. The Institute would not just be an ivory tower such as traditional universities. Its faculty would be uniquely positioned to address the economic issues that were afflicting the country and the world.[7]

On Flexner's trip to Princeton the humanities and mathematics received particular attention. By then his routine was well established. The individual visits allowed Flexner to engage in some public relations work while obtaining guidance. He would offer an Institute Bulletin and seek advice on general organizational issues. Flexner then posed the problem of how best to attack the consultant's subject. The informal discussions naturally led to personnel recommendations. These were often elaborated in a follow-up memo. Through this process Flexner accumulated a list of candidates for faculty positions.[8]

Flexner himself had a personal interest in art history and classical studies. His Princeton rounds yielded some names in these areas, but he failed to connect with Charles Morey, who was chair of the Department of Art and Archaeology. Morey would later emerge as one of Flexner's most influential advisors. Even if they had met at this time, it is unlikely that humanities would have moved to the front of the line. Flexner was already acquainted with some of the candidates that Morey would propose.[9]

Among the Princeton mathematicians that Flexner consulted was Solomon Lefschetz. Lefschetz was noted both for his powerful conceptual insights and for his directness. The advice to Flexner on where to begin was characteristic:

> Annex . . . the younger group of geometers. It is the most vital and promising of mathematical groups in the U.S., the one with the highest national and international standing. It includes Veblen and Alexander of Princeton, Birkhoff and Morse of Harvard and also myself . . . Hermann Weyl is the only mathematician anywhere definitely above these names. But as he occupies the most distinguished mathematical chair in the world (in Göttingen) I do not see him giving it up.[10]

Lefschetz's comments resonated with Flexner's aspirations. Geometry might be a promising area to make a start. Birkhoff, who had also been mentioned by Smith, was known to Flexner as an influential advisor for the International Education Board of the Rockefeller Foundation. While Flexner did not contemplate such a large group, he was covertly considering Princeton as a possible location. If he could secure Birkhoff, then, together with the group already established in Princeton, the Institute could make an international statement in geometry. [11]

Flexner had set out to identify the best American scholars in each field. In mathematics at that time the names that he was most likely to hear were Leonard Dickson of Chicago, G. D. Birkhoff of Harvard, and Oswald Veblen of Princeton. Each had been a student of E. H. Moore at Chicago and gone on to an exceptional career of scholarship. They were the only Americans of their generation who had delivered plenary addresses to the International Mathematical Congress. A further consideration for Flexner was that each had demonstrable success as a thesis advisor.

Chicago and Harvard, the final two destinations for the American portion of Flexner's survey, coincided with the home bases of Dickson and Birkhoff. Mathematics remained one of several initiatives under consideration. Flexner's correspondence does not reveal the impression that Dickson made on him when they met in Chicago. Whether or not Flexner identified the genius that he was seeking, there were objective criteria that mitigated against selection of the Chicago mathematician. Dickson was an algebraist who would not have contributed to a program in geometry. Age may have been a factor as well. Dickson was then 57. With the Institute's opening a few years in the future, Flexner preferred someone younger.[12]

One frequent criticism of the Hopkins mathematics department was that it had been too dependent on the elderly Sylvester. Despite the miracles that he performed over eight years, there was a precipitous decline after his departure. Some argued that a department should not rely on a single individual for its vitality. Flexner was well acquainted with this analysis. Given his own age, Flexner anticipated a brief tenure as Director. The second round of appointments would be made by his successor. If he could only make one appointment in a department, it was important that the Flexner faculty remain in place sufficiently long to establish a solid tradition.

In his interviews with prospective faculty Flexner had a number of objectives. In addition to the consultation aspect, Flexner needed to judge whether the individual had the right stuff to serve on his faculty. If this were the case the next steps were especially delicate. Flexner was not yet in a position to offer an appointment. He first intended to seek Bamberger's approval and perform some groundwork with the Board. It would be embarrassing to follow this course and then have an offer declined. Ascertaining whether an individual might accept an unoffered position required diplomacy and the intuition to send and receive coded signals. Even if he were unable to evaluate the importance of a mathematical result, Flexner was confident in his ability to surmount these obstacles.

At Harvard Flexner was finally to meet the 46-year-old Birkhoff. The highlight of his Cambridge visit, however, was a long conversation with law professor Felix Frankfurter. The future Supreme Court Justice possessed a broad knowledge of the social sciences. Frankfurter was enthusiastic about the Institute's potential to elevate academic scholarship in the United States. Flexner came away energized that his vision was shared by such a highly regarded intellectual. It remained to define a further role that Frankfurter could play in launching the Institute.[13]

The meeting with Birkhoff was productive, but inconclusive. His views on program design were compatible with those of Flexner. Birkhoff proposed that Flexner "secure permanently one or two mathematicians of great and undisputed genius."[14] There should also be younger staff members who would eventually move on to positions at other universities. Birkhoff fully approved of concentrating on selected subspecialties, rather than attempting to cover all of mathematics.

Birkhoff must have understood that he was under consideration to lead mathematics. As both parties sized up the other, the dialogue was cautious. All that Flexner was able to gauge was that Birkhoff was neither willing to commit nor disqualify himself. They parted agreeably, planning a follow-up meeting for April in Paris. With Birkhoff established as the putative choice in mathematics, the feasibility of this subject hinged largely on his availability.

George David Birkhoff was of Dutch descent. He was born in Overisel, Michigan, on March 21, 1884. His interest and talent in mathematics became apparent during adolescence. Birkhoff received the best mathematical training available in the United States at that time. He was an undergraduate at the University of Chicago and then transferred to Harvard. There he began research under the influence of Maxime Bôcher, himself a student of Felix Klein.[15]

For his PhD, Birkhoff returned to Chicago, where he completed his thesis in 1907 under the direction of E. H. Moore. The topic was asymptotic expansions of linear ordinary differential equations. His techniques exhibited the classical analysis approach of Bôcher and the more modern influence of Moore. Combining his mathematical contacts at Harvard and Chicago, Birkhoff was already networked with the American mathematical elite for the present and future decades.

Birkhoff's first position was a $1,000 per year instructorship at the University of Wisconsin. Within a short time other universities began to approach him. In particular, Princeton had recently initiated an effort to upgrade its mathematics faculty. This began in 1905 when its president, Woodrow Wilson, established new junior faculty positions known as preceptorships. Wilson was attempting to enhance the undergraduate program by decreasing class size and increasing faculty-student interaction. Under the leadership of H. B. Fine, another Klein student, the mathematics department filled its preceptorships with exceptionally promising young scholars. Oswald Veblen was among the first wave. It was a few years later, in Birkhoff's second year at Wisconsin, that Fine offered him a Princeton preceptorship.[16]

Veblen and Birkhoff were friends from the time they overlapped at Chicago. Veblen encouraged Birkhoff to join him in Princeton. Birkhoff had recently married and was hoping to improve his circumstances at Wisconsin. Fine's patience was tested as Birkhoff delayed his decision while waiting to see how the slow-acting state university would respond. Birkhoff finally agreed to the Princeton terms, but not before the Chicago faculty joined Fine in questioning his dilatory handling of the situation.[17]

Birkhoff came to Princeton as a preceptor at a salary of $2,000. This period was marked by stimulating interaction with Veblen. Birkhoff had become interested in the recent work of Henri Poincaré. Veblen recalled,

I remember well how frequently, in the walks we used to take together during his sojourn in Princeton, Birkhoff used to refer to his reading in Poincaré's "Les Méthodes Nouvelles de la Mécanique Céleste," and I know that he was intensively studying all of Poincaré's work on dynamics.[18]

Later, Veblen would adopt another geometric technique of Poincaré.

Meanwhile Fine's appointments were maturing into the mathematical leaders of the future. Within a few years Princeton joined Harvard and Chicago as the most highly regarded American mathematics departments. Accompanying the recognition of their faculty's outstanding scholarship was competition for their services. By Birkhoff's arrival in 1909, two of the original four preceptors had left for other opportunities. Another was promoted to professor and Veblen was about to receive an offer from Yale. Fine successfully moved to retain Veblen by effecting his promotion as well.

Harvard was then the foremost mathematics department in America. It was led by the European-trained analysts Bôcher and William Osgood. Both had taught Birkhoff during his undergraduate days at Harvard. They had followed his subsequent career with considerable interest. For Harvard to remain on top they needed to identify the best rising talent. Bôcher, in particular, saw in Birkhoff a young mathematician of exceptional skill and taste. In 1911 Harvard offered Birkhoff an assistant professorship.

Fine countered with a promotion and raise in salary. Thus Birkhoff remained at Princeton as a 27-year-old professor at $3,500. During the negotiations a significant bequest became available to Princeton. Members of the mathematics department believed that these funds would provide them with additional positions. When the administration elected to allocate the resources in another direction, Birkhoff informed Bôcher of his desire for renewed consideration.[19]

In 1912 Harvard again offered Birkhoff an assistant professorship. The salary was $2,500, still $1,000 below his Princeton salary at the higher rank. Moreover he was slated to remain an assistant professor for at least seven years, albeit with a raise to $3,000 after two years. Birkhoff accepted the Harvard offer. Fine, who had no understanding that he had been under probation to secure the bequest for mathematics, was flabbergasted. While Birkhoff's motivations may have been complex, it is clear that he was not simply seeking to maximize his income.[20]

Birkhoff immediately validated the competition for his services. He proved a huge conjecture of Poincaré. From the late nineteenth century until his death in 1912, Poincaré was regarded as the greatest mathematician in the world. In his work on celestial mechanics Poincaré was stymied by a problem involving functions mapping an annulus into itself. In his last years Poincaré had publicized this problem in the hopes that someone else might solve it. It became known as Poincaré's Geometric Theorem and was the object of intense interest throughout Europe.[21]

To state the problem let the annulus A consist of the region in the plane bounded by two concentric circles. Consider a function f that assigns each point of A to another point (possibly itself) and satisfies the following stipulations: The function is continuous (nearby points of A are taken by f to other points that are nearby to each other) and area preserving (subregions of A are taken to other subregions with the same area). Furthermore the points on one boundary circle are sent to other of its points in a clockwise manner while the second boundary circle is counterclockwise. Poincaré conjectured that for any such function there must be at least two points of the annulus that were fixed by f (that is, there exist w and z in A with $f(w) = w$ and $f(z) = z$).

When the 28-year-old Birkhoff discovered a proof, it was a profound event in the world of mathematics. Not only was Birkhoff the subject's newest star, but the United States finally gained respect in Paris, Göttingen, and the other established mathematical centers. By some accounts, Birkhoff's result marks the coming of age of the United States in mathematics.[22] Curiously enough it did not hasten Birkhoff's advance through the ranks at Harvard. He did not become a professor until 1919, one year after his induction into the National Academy of Sciences.

Birkhoff's career flourished at Harvard with groundbreaking work on dynamical systems. Recurrence and Lebesgue measure were fruitful threads that he wove into the theory. Among the phenomena he identified were elements of a notion that later became known as chaos. Throughout the twenties he received prestigious awards, honorary degrees, and recognition as a foreign member of various European scientific societies. Birkhoff's accomplishments, age, and standing presented an ideal profile for an Institute professor.

When Flexner completed the domestic portion of his survey, he had narrowed the opening subject down to either mathematics or economics. To lead economics Flexner was considering Jacob Viner. Viner was a progenitor of what would later become known as "the Chicago School" of Economics. The Canadian-born Viner was eight years younger than Birkhoff. Viner, who became a United States citizen in 1924, received his PhD at Harvard and then joined the faculty at the University of Chicago. Highly regarded for both his

Jacob Viner. (Courtesy of Ellen Seiler.)

depth and breadth, Viner was an authority on international economics who was well versed in theory and history. His Chicago graduate course on economic theory was legendary. Some of the twentieth century's most celebrated economists, such as Milton Friedman, Paul Samuelson, and George Stigler, vividly recall both a seminal learning experience and the terror evoked by Viner's Socratic approach. Among Viner's advocates to Flexner were Beard and Frankfurter.[23]

Personnel was the crucial element of Flexner's decision between Viner/ economics and Birkhoff/mathematics. The real showdown was shaping up for April in Paris, where Flexner would interview both Viner and Birkhoff. Flexner's trip to the continent had already been programmed in order to gain the perspective of European intellectuals. Flexner reached England early in March, 1931. The arrival was timed to coincide with a vacation break for his daughter Eleanor, who was studying at Oxford. Abe was concerned about Eleanor's health. The trip afforded him the opportunity to spend some time with his daughter and make a personal assessment of her condition. After some discussions in Oxford and London, the Flexners departed for Rome. It was a destination that was suggested by Princeton mathematicians. Sightseeing in the warmer clime had the added benefit of aiding Eleanor's recovery from the British winter.[24]

Flexner reported that his consultations with Roman mathematicians reinforced the opinions he had already formed. Most likely this included strong support for Birkhoff, Veblen, and Lefschetz. Lefschetz' work in algebraic geometry had profoundly advanced a program begun in Italy. Three years earlier Birkhoff and Veblen were the only Americans to deliver plenary addresses at the International Mathematical Congress when the quadrennial meeting was held in Bologna. Birkhoff's work was especially prominent as it had been in an Italian journal that Poincaré had posed his Geometric Theorem. [25]

From Italy, Paris was the next stop on Flexner's itinerary. There he would see Jacques Hadamard, another enthusiast of Birkhoff. Flexner, who revered European scholarship, understood Hadamard to be the leading mathematician of France. Hadamard's timely buildup of Birkhoff carried considerable weight.

Paris was the setting for the crucial interviews with Viner and Birkhoff. With these meetings just a few days apart, Flexner had an ideal opportunity to take their measure and compare. On another level, he himself was being tested to exercise the type of judgment that was the hallmark of Daniel Coit Gilman. Flexner's reactions are depicted in the following excerpts from letters to his wife and daughter.

> Viner came from Geneva yesterday: not highly polished—but very able, I think: We talked from 9 until two and I got what I wanted.[26]

> Yesterday, Birkhoff, the Harvard mathematician, probably our best, had tea with me and I blazed forth: I had seen him before leaving home; but the thing has become more vivid and real to me, and I could see that he, too, was more infatuated.[27]

> Since I last wrote I have had delightful talks with Hadamard, first of living French mathematicians; it seems common opinion that Birkhoff of Harvard is not only our best, but among the best anywhere. Fortunately he arrived here to lecture at the College de France Thursday: his wife and he had tea with me yesterday. My guess is—tho' we skirted the problem—that he would jump at the job. He thoroughly approved the scheme as it now lies in my mind, "schools"—as in Rome and Paris—of various subjects: a head, some younger men for 3–5 years at first, and a constant stream of great leaders from elsewhere for a whole year at least. Hadamard said: "That would be Heaven." "Would you come sometime?" I asked. "Just give me the chance."[28]

Despite the superb credentials of both candidates, in his interviews Flexner divined a marked difference. He had made his decision on the first program for the Institute. Flexner wanted the package of mathematics and Birkhoff. If his impressions of Viner and Birkhoff had been reversed, it is likely that economics would have won out.

The details were subject to change, but, at last, Flexner had a scenario and a structure. The Institute would begin with a mathematics "school" headed by Birkhoff. The staff would consist of the most promising new or recent PhDs and feature yearly visitors such as Hadamard and Hermann Weyl. With no undergraduates and a small number of graduate students, teaching duties would be minimal. It was a formula for a world-class mathematical research environment, unprecedented in the United States. The headliners Birkhoff and Hadamard were certain to attract the best students in the country. The model could be replicated in other subjects.

Much would continue to evolve, but Flexner never looked back from his choice of mathematics. It was a relief to have such a central issue resolved and Flexner traveled through Germany in high spirits. With stops in Berlin, Dresden, Cologne, Bonn, and Hamburg, Flexner conferred with old friends and other authorities. It is notable that he bypassed Göttingen, apparently confident that mathematics was under control.

Economics remained another matter. In Bonn Flexner had a long meeting with Joseph Schumpeter, a prominent Austrian economist. Schumpeter was a well-connected scholar who quickly won over Flexner by effusively praising the Institute concept. The economist's strongest recommendations were for Ragnar Frisch of Norway and 25-year-old Wassily Leontief of Russia. Frisch would win the Nobel Economics Prize when it was first awarded in 1969, and Leontief received it four years later. Although Schumpeter rated Viner highly, and first among Americans, Flexner continued to have doubts. One significant development did occur in Flexner's thinking on personnel. He penciled in Schumpeter as a one-year visitor for the start-up of the economics school.[29]

Flexner completed his loop of European travel by returning to England. There he reunited with Eleanor who, earlier, had resumed her schoolwork. Another round of interviews in Cambridge, Oxford, and London preceded the cruise for home. Flexner left with the satisfaction that crucial issues had been resolved and that the Institute was on the right course.

There remained one huge obstacle. Flexner must have dreaded the topic, knowing that it would inevitably arise in each consultation. Where would the Institute be located? Not only did Flexner lack a satisfactory answer, but he felt constrained in even discussing possibilities such as Princeton. From the outset Bamberger and Fuld were unequivocal in their intention to fund a project with a Newark area location. It was their way of giving back to the community that they cherished. In fact, they were exercising little other influence over their magnificent benefaction. The Institute bore a generic name. Flexner was given considerable latitude in program design and personnel selection. From the perspective of Bamberger and Fuld there was no location issue to discuss, except whether to locate on their estate or procure a nearby site.

While Flexner fully understood the sentiments of Bamberger and Fuld, he also recognized a once-in-a-lifetime opportunity. All of his previous efforts had failed to bring about a true graduate university in the spirit of Gilman. It was unlikely that providence was going to connect him with another philanthropist willing to commit the necessary funding. Only Bamberger and Fuld could fulfill Flexner's dream. The difficulty was that Newark was a poison pill. Flexner needed to recruit the world's greatest intellectuals. These were scholars who were already part of the communities at Harvard, Princeton, and Chicago. Newark could not compete. Moreover, it was essential to provide a first-class library. Flexner had no intention of investing in this expensive resource. The only solution was to locate near a satisfactory existing library, or, equivalently, a university. The absence of a university made Newark an unsuitable venue for either a medical school or the Institute for Advanced Study.

Flexner was confident that, over time, he could persuade the founders to relent on Newark. He must have gingerly raised the issue with Bamberger at an early stage, but found that it was treacherous ground. Officially, Flexner treated location as a matter to be deferred while he concentrated on personnel and structure. His most revealing move was to slip the general language about New Jersey past Bamberger and into the letter to trustees. At the very least this provided an avenue in the event of Bamberger's death.

Tracing Flexner's other moves on the location issue is difficult. He could not take any chance of word reaching Bamberger. In June, 1931, following his return from Europe, Flexner began the campaign for Princeton. It is likely that he settled on the New Jersey venue earlier, possibly from the start. Almost certainly he had explored its feasibility in his Princeton trip at the beginning of the year.

Flexner was aware that the notion of a mathematics institute in Princeton had been pushed by Oswald Veblen for many years. Veblen believed that mathematical research was inadequately supported by universities. Serious researchers were diverted from their efforts by heavy teaching loads and service demands. It was a view remarkably compatible with that of Flexner. Beginning in the mid-1920s Veblen set out to improve the opportunities for mathematical research in America.

In 1923 Veblen successfully lobbied to include mathematics among the scientific subjects in which the National Research Council awarded postdoctoral fellowships. This afforded selected young mathematicians an opportunity to establish a research program at a crucial stage of their career. It also brought Veblen into contact with Simon Flexner who was already involved in administration of the scientific awards.

Veblen next proposed the establishment of a mathematics research institute at Princeton. At an early stage he submitted his ideas to the National Research Council and to Simon Flexner. Flexner referred Veblen to his brother Abe at the General Education Board. Veblen's pitch was among the many that Abe received in 1924. Under Veblen's plan an outside agency would fund the salaries of several senior faculty and an equal number of junior people, releasing all of them from normal teaching duties. The university would supply office space and library access. Veblen predicted that this structure would create a fertile culture for mathematical research that attracted National Research Fellows and graduate students. Veblen's plan was similar to the vision conceived by Flexner in Paris.[30]

Joining Veblen in the proposal were two influential mathematics colleagues. H. B. Fine and L. P. Eisenhart were respectively deans of the Graduate School and Faculty. Princeton University did submit the institute proposal to the General Education Board in 1925. For Veblen, the outcome was mixed. The General Education Board funded a $1 million challenge grant for which the university was required to raise $2 million in matching funds. The money was to establish an endowment to support research in all the sciences. In particular, several research chairs were created. Veblen did not get his institute, but he became the first H. B. Fine Research Professor of Mathematics.[31]

A considerable portion of Princeton's matching share was provided by alumnus Thomas Jones, a wealthy friend and former classmate of Fine. Veblen had not given up on his research institute, but no prospect for funding was in sight. When Fine died in 1928 from injuries sustained in a bicycle accident, the mathematics department received an unanticipated largess. As a tribute to Fine, Jones underwrote the cost of a new mathematics building. No expense was to be spared in the construction of Fine Hall.[32]

Fine Hall was not a research institute, but it was a significant step toward realizing Veblen's cooperative environment for mathematical research.

Among its revolutionary steps were to provide each faculty member an office, establish a mathematics library, and create gathering places for the discussion of mathematics. The lavish appointments, selected by Veblen, were on a scale that was unprecedented in academic life. From its opening in 1931, Fine Hall had a profound impact on the international mathematical scene. For decades, it would remain a topic of fond recollections.[33] Fine Hall also gave Princeton an edge in its competition with Harvard for recognition as the leading American mathematics department.

The thread of contact between Veblen and Flexner runs throughout the creation years of the Institute for Advanced Study. Even as Flexner was drafting the Bamberger-Fuld codicil, in January, 1930, he was corresponding with Veblen on a related subject. Flexner first responded to mention of a Veblen speech in the *New York Times*. Veblen had deplored the low standing of research in American universities. This initiated an exchange in which the two men agreed on the need for establishing a "seat of learning," each thinking of his own institute as the solution.[34]

Several months later Veblen read the *New York Times* announcement of the Institute for Advanced Study. He was quick to send a congratulatory note to Flexner. Veblen took the occasion to propose Princeton as a suitable site. The next correspondence on record was in June, 1931. The context of this letter, written by Veblen, suggests that it is in response to questions raised at a recent meeting. This places Flexner in Princeton after his return from Europe.[35]

With the selection of mathematics as the Institute's first school, there were compelling advantages to a Princeton venue. Assuming the university were willing, the Institute could temporarily rent space in Fine Hall. Not only would it be unnecessary to construct a new building, but, together with the Princeton mathematicians, the Institute would create a world-class seat of mathematics in the United States. To explore a cooperative agreement with the Princeton authorities, Flexner would have dealt with Dean Eisenhart and President John Grier Hibben.

It was a sensitive negotiation and Flexner was operating from a weak position. He was seeking to utilize Princeton's valuable infrastructure, but to remain completely autonomous. What he was offering in return was a vision of future shared glory. Many administrators would have insisted on some control over Flexner's campus operation. Eisenhart, however, had an unusually broad perspective. He was dedicated to advancing mathematical scholarship. Eisenhart and Flexner would develop a warm friendship of mutual respect.

By June 1931, Flexner must have received some indication that Princeton was receptive to hosting the Institute under his terms.

It remained to overcome Bamberger's insistence on Newark. Bound by his policy of avoiding confrontation with the founders, Flexner turned to selected trustees for intercession. Aydelotte tried first, reporting that he was unable to dissuade Bamberger and Fuld from seeing their estate as the site for the Institute. Flexner welcomed his efforts while stressing his disassociation from the campaign. Flexner then appealed to Maass and Leidesdorf. When Maass appeared sympathetic, Flexner was emboldened to rehearse the following argument in an intense week-long correspondence: The Bamberger-Fuld letter to trustees had stipulated that the Institute "be located in the State of New Jersey." While the by-laws used the language "at or in the vicinity of Newark," the letter was the prevailing expression of intent. Under these circumstances the central New Jersey town of Princeton could well be interpreted as in the vicinity of Newark. The trustees could simply choose Princeton.[36]

Maass refused to be drawn into the ploy. He maintained that to locate in Princeton would require Bamberger's explicit consent. Flexner's agenda is ambiguous. On the surface he appears to prepare the groundwork for a trustee move against the founders. This was inconceivable. A more likely interpretation is that Flexner was manipulating Maass to use his own influence with Bamberger. The attempt failed.[37]

Flexner remained in New York for the month of June, regularly lunching with Bamberger, Maass, and Leidesdorf. It is unclear whether there was any vetting of the proposed Princeton solution, but Flexner was feeling the pressure from his consultants. After all, he could not expect Birkhoff, or anyone else, to accept an offer for a "to be arranged location," especially with the benefactor ordering stationary for Newark. Flexner remained patient. As tempting as it was to seize the opportunity and sign up an eager Birkhoff, protocol demanded some sort of clearance from the trustees.[38]

With the next Board meeting scheduled for October, there was little to be gained by pushing Bamberger on location. In July, Flexner departed for his annual summer retreat in the Canadian woods. There he would prepare a detailed report so that the trustees could endorse an initiative in mathematics. Among the recommendations would be the appointment, by Bamberger, of a committee to examine the site questions. Aydelotte and Maass were certain to be selected as members. It would then be their duty to weigh the merits of Princeton. The pragmatic, business side of Bamberger could not discount the advice of a committee that he himself had commissioned.

In his 20-page confidential document to the trustees, Flexner presented his most detailed vision of the Institute for Advanced Study. He began with a reaffirmation and expansion of the parameters that had been established one year earlier:

> the Institute for Advanced Study should be small, that its staff and students or scholars should be few, that administration should be inconspicuous, inexpensive, subordinate, that members of the teaching staff, while free from the waste of time involved in administrative work, should freely participate in decisions involving the character, quality, and direction of its activities, that living conditions should represent a marked improvement over contemporary academic conditions in America, that its subjects should be fundamental in character, and that it should develop gradually . . . . The Institute for Advanced Study will be neither a current university, struggling with diverse tasks and many students, nor a research institute, devoted solely to the solution of problems. It may be pictured as a wedge inserted between the two—a small university, in which a limited amount of teaching and a liberal amount of research are both to be found.[39]

The basic organizational unit was to be the "school," deliberately distinguishing it from the traditional department of university life. Flexner envisioned these schools as flexible and varying over time (the word he often employed was "plastic"). For example, the school of mathematics might open with an emphasis on geometry, but at some future time it could shift to algebra or even be dissolved. Philosophy of science may or may not be among its areas of interest. Each school was to select its own students. Neither admission nor degree requirements were necessary. Such mandates introduced needless bureaucracy and interfered with the function of a wise faculty.

Several of Flexner's operational axioms should be noted, as they impinged on his future actions. One was derived from his experience in medical education. Both students and faculty were to be engaged on a full-time basis. This was the element that he had introduced with Welch that was so vital to the modernization of American medical education. Flexner recognized full-time immersion as an essential ingredient of productive research. With the salaries he contemplated, it would be unnecessary for professors to prostitute themselves to textbook writing or other part-time duties.

Flexner included two sound management principles. No collaboration was to be forced. It could only develop naturally among the partners. With regard to the budget, Flexner cautioned against deficit financing. He had seen its disastrous effect on universities. Of course it was easy to insist on a surplus as he opened with an unencumbered $5 million endowment. The challenge would occur in the future when opportunities for expansion had to be balanced against limited funds. In handling this dilemma Flexner would find that it was easier to criticize than to have responsibility for making things work.

One fourth of the document was devoted to Flexner's thinking on subjects and schools, without naming any prospective professor.

> I assume at the outset that no subject will be chosen or continued unless the right man or men can be found. Subject to this reservation, never to be forgotten, a very vague statement is contained in Bulletin No. 1. I can be somewhat more definite now, though retaining liberty to change up to the very moment when action is resolved upon. The decision not to begin with the physical or biological sciences has become stronger; moreover, they are creating problems with which universities are not now dealing competently. Finally, they are not at the very foundation of modern science. That foundation is mathematics; and it happens that mathematics is not a subject in which at present many American universities are eminent. Mathematics is the severest of all disciplines, antecedent, on the one hand to science, on the other, to philosophy and economics and thus to other social disciplines.

As further justification for his choice of initial subject Flexner noted,

> it is no small, though an accidental and incidental advantage, at a time when we wish to retain plasticity and postpone acts and decisions that will bind us, that mathematics is the simplest of subjects to begin with. It requires little—a few men, a few students, a few rooms, books, blackboard, chalk, paper, and pencils.

Later, in his autobiography, Flexner repeated that he chose mathematics because it was fundamental and required little initial investment, but also added a third reason: "I could secure greater agreement upon personnel in the field of mathematics than in any other subject."[40]

Both fundamentalism and economy were attractive features of mathematics. Neither, however, was the driving force behind its selection. Philosophical arguments were mere justifications to impress trustees, while thrift might become an issue in selecting a third or fourth program. For his first school Flexner needed to demonstrate that the Institute was a new force in international scholarship. He had indeed found considerable agreement in appraising the quality of mathematics and mathematicians in the United States. Princeton and Harvard were the only two departments with solid international standing. Both Chicago and Hopkins had slipped. As to individuals,

Birkhoff was the sole outstanding American mathematician. Flexner's decision to begin with mathematics was a consequence of the following combination of factors: Birkhoff's eminence, Flexner's personal impression of Birkhoff, the likelihood of Birkhoff's participation, and the prospect of bringing Birkhoff to Princeton to form a super-group that would stand among the leading mathematics departments in the world.[41]

Flexner portrayed the school of mathematics as laying a foundation for his next program. "Assuming that funds are adequate and that the right persons can be secured, I am now inclined to include economics." Flexner was hopeful that economics would appeal to the large business faction on the Board. Unlike mathematics, it was an accessible subject that could be motivated in starkly practical terms. Flexner was not bashful about raising expectations:

> Time was, when Europe was exposed to ravage by typhus or bubonic plague. Their origin and progress were shrouded in mystery; but the veil has now been lifted; these plagues will not recur, because their causes and methods of distribution are understood; they can be prevented or stopped. But from social and economic plagues the world is not yet immune. They continue to come and go mysteriously. We cannot any longer sit helpless before these social and economic plagues, which, once well under way, ravage the world, as our present economic and social perplexities and sufferings show. The very conquests which science has wrought—increased production and easier distribution, which ought to be blessings—have drawn in their wake curses that may or may not be connected with them. On these intricate and recondite matters I have no opinion; but clear it is that nowhere in the world does the subject of economics enjoy the attention that it deserves—economics in the broad sense, inclusive of political theory, ethics, and other subjects that are involved therein. The Institute for Advanced Study has here a pressing opportunity; assuredly at no time in the world's history have phenomena more important to study presented themselves. For the plague is upon us, and one cannot well study plagues after they have run their course; for with the progress of time it is increasingly difficult to recover data, and memory is, alas, short and treacherous.

Flexner had watched his own family suffer through the panics of 1873 and 1893. The current "panic" was emerging as a global event of great magnitude. Paul Warburg had given Flexner hope that these economic problems were tractable. Warburg, who was soon to have a stroke, could offer no additional guidance. Warburg's death left Flexner with an ambitious objective for the economics school. He still had a distinguished stable of social science advisors that included Charles Beard, Felix Frankfurter, and Joseph Schumpeter. Reconciling these and other powerful, disparate voices into a coherent program would be a continual struggle.

Flexner approached mathematics in a markedly different manner from economics. With mathematics Flexner had no hope of understanding the substance of problems or approaches. He learned that there was an area called geometry and that the names of Birkhoff and Veblen were held in high esteem. Flexner's personal assessment of these men was largely based on social impressions. In the School of Mathematics professors would have complete freedom to pursue their research without scrutiny or interference.

Flexner's report went no further than nominating mathematics and economics. He did not suggest a third program, nor did he divulge any of the names he was contemplating for faculty appointments. There were several other issues on which the report was deliberately vague. Flexner was no longer wedded to the personnel structure that he and Birkhoff had contemplated in Paris. This was, in part, due to the interdisciplinary demands of the program he was developing in economics. There he required expertise in history, political science and possibly other subjects. Perhaps more than one senior faculty member would be needed.

Another cloudy matter was the foreign/domestic mix. Flexner had a great admiration for European scholarship. He wanted to create Institute schools with a European influence and a distinct American character. His earlier model was to secure an American head and supplement with a robust flow of one year visits by foreign luminaries such as Hadamard and Schumpeter. Despite the Institute's founding ideals of inclusiveness, Flexner was hesitant to appoint Europeans to permanent faculty positions. As the talent search continued, it was becoming difficult to reconcile this policy with his aspirations for world-class Institute schools.

One of the thorniest problems involved defining roles and relations among the faculty, trustees, and director. Flexner often ridiculed university boards of businessmen who were out of touch with crucial academic matters. At the other extreme he had observed difficulties in the British system where the faculty was vested with control. Moreover, he was determined to protect the Institute professors from being engulfed in administrative duties and infighting. Flexner cautioned that relying on the Director as the bridge between the faculty and trustees might tend to produce a leader who was "autocratic and unlikely to be informed."

As a provisional solution Flexner's report urged that the Board of Trustees consist of lay-people, academicians from other institutions, and members of the Institute faculty. Even if it did not guarantee an atmosphere of civility, it would provide a forum for all relevant points of view. Flexner had ample reason to anticipate controversy for his tripartite concept. Consultants such as Frankfurter and Veblen had argued vigorously against lay-members and Simon continued to raise the issue of budget control. Bamberger had resisted the original attempt at a more scholarly board. Now he and his associate Percy Straus were skeptical of including workers among the management. With all fifteen board slots occupied it could be argued that there were no vacancies for the first faculty.[42]

Flexner presented the report to the Board at its October meeting. He requested continued study, but no formal action, excepting the authorization for Bamberger to appoint a Committee on Site. The trustees were taking their cues from Bamberger and giving Flexner free rein. The choreography of personnel moves remained hostage to resolving the location problem. Flexner planned to seek approval at the January meeting to go forward with the two schools. By that time the site issue should be settled. Flexner would then recruit his first faculty. The offers would be preapproved by Bamberger and be made contingent on ratification by the Board. By this design, the trustees would only learn of names that were already committed.

With his first offers still a few months away, it was important to keep the pot boiling with Birkhoff and Viner. A speaking engagement in Cambridge gave Flexner an opportunity to update Birkhoff. Flexner disclosed that the list of possible locations had been narrowed to Princeton and Washington. By sharing a copy of his report to the trustees and soliciting a reaction, Flexner was essentially previewing their negotiation.

Birkhoff's follow-up was revealing. He embraced the report and the choice of subjects. On the issue of personnel Birkhoff wrote,

> The Institute must therefore secure men of the highest possible calibre, outstanding figures of their day . . . . In order to get this initial staff it is obviously necessary to set your maximum salaries at a rather high level. I think much will depend upon the decision of the Institute in this direction.[43]

As to possible locations, Birkhoff expressed approval and preference for Princeton. These responses bolstered Flexner's confidence that Birkhoff would react favorably to an offer of an above-market salary at a Princeton Institute.

A few weeks later Flexner happened to be in Chicago. He was still waffling over whether he wanted Viner, the University of Chicago economist.

Viner was an attractive candidate. He had enthusiastic backing from Schumpeter, Frankfurter, and others. If the School of Economics began with Viner,
it would have a 40-year-old leader who was widely regarded as the best economist in America. Given Flexner's demographic preferences, it is difficult
to imagine a better fit. In Chicago, Flexner and Viner met for the second
time. After another long conversation, Flexner remained uncertain over how
he would staff economics.[44]

Flexner was making progress behind the scenes on the site issues. This was
indicated by his assurance to Birkhoff of the exclusion of Newark. Meanwhile,
Bamberger was just beginning to select the Committee on Site. His slate
consisted of Maass, Aydelotte, Edgar Bamberger, Flexner, and the founders.
Flexner humbly suggested that the committee would be strengthened with
Weed replacing himself, pointing out that as Director he was permitted an ex-
officio participation. Bamberger accepted the amendment and the committee
was formalized with Maass as chair.[45]

Flexner had skillfully manipulated the staffing. To counteract the Bamberger family faction were Maass, Aydelotte, and Weed. Maass and Aydelotte
had already been won over to the Princeton side. Whether or not Flexner was
informed of the opinionated Weed's predisposition, which is likely, it was inconceivable that the Dean of the Johns Hopkins Medical School would have
favored Newark over Princeton. While such committees acted by consensus
rather than majority, the three-to-three division and his ex-officio standing
would call for Flexner to act as a mediator, rather than a partisan.

The first site committee meeting took place in early December. The nomination of Washington had been made by a number of consultants. Lefschetz
had pointed out that its Federal standing made it a part of every state, including New Jersey. Others appreciated the value of the Library of Congress as
a resource. There is no evidence that Flexner ever gave serious consideration
to Washington, but it was desirable to have a third candidate that diluted the
appearance of a Newark-Princeton showdown.

The primary outcome of the committee meeting was implementation of a
suggestion by Flexner to solicit advice from a number of scholars. The Director was commissioned to produce the letter draft and mailing list. The letter
was another step in Flexner's subtle campaign. He refrained from calling a
vote on Princeton and Newark, for which the outcome would be predictable
and offensive. No cities were listed, nor was there a specific charge for the
respondents to name a favored location. Rather, Flexner requested a description of the conditions that would best facilitate the operations of the Institute.
Responses, of varying detail, would arrive over the next few months.[46]

On the day following the site committee meeting, Flexner undertook a diplomatic mission to Princeton. The record does not reveal his charge, if any, from the committee. Presumably he was firming up arrangements with the university in the likely event that the Princeton venue were selected. At the top of his list would have been temporary mathematics quarters in Fine Hall and library access for all Institute scholars. While Dean Eisenhart and President Hibben were the official representatives of the University, Veblen's support was a valuable asset.

When Flexner and Veblen met in Princeton both men recognized that there was common ground in their views on research and universities. At this particular meeting, however, each was holding information that was unknown and of considerable interest to the other. That they elected to pool these resources elevated their relationship to a partnership of trust and dependence that would have enormous ramifications on the direction of the Institute for Advanced Study. Flexner would disclose that he planned to open the Institute with a School of Mathematics headed by Birkhoff. Veblen would reveal that Hermann Weyl, who had given up a Princeton professorship two years earlier, might be interested in returning to the city.[47]

Veblen had recently received an invitation from Weyl to deliver a series of lectures at Göttingen. In the letter Weyl alluded to his fondness for Princeton and his regret over taking a position at Göttingen. Veblen had already replied, suggesting that, if he were serious, other opportunities might become available. Flexner had to be intrigued at the prospect of procuring the scholar whom his survey had revealed was the most coveted mathematical figure in the world. Meanwhile Veblen was beginning to realize his own long-standing ambitions for a mathematical research environment in Princeton. [48]

Prior to their parting, Veblen drove Flexner around the surrounding area, pointing out parcels of land that might provide a desirable setting for the Institute campus.[49] On the next day Flexner followed up by sending Veblen a copy of the confidential report to the trustees. Veblen responded enthusiastically, including the following paragraph that should be read in the light that he had a better understanding than Flexner of Birkhoff's predilections and previous practices in dealing with job offers:

> You intimate that you would not go ahead in a particular field if you were not able to get the "the right man". My belief is that in most fields, there are sufficiently many good men so that you can surely get a man of the right sort. For example in mathematics, if you cannot secure the man whom you have picked out and whom I agree is the best first choice, there are a number of others who are surely as good and who may, in fact, be better.[50]

Despite the absence of names, it was not a hypothetical discussion. Birkhoff was the first choice. Flexner's selection of mathematics had been predicated on the availability of Birkhoff. Veblen knew, regardless of the impression that Flexner had received from Birkhoff, that there was a substantial likelihood that Harvard would succeed in retaining its foremost mathematician. If Birkhoff did decline then Veblen was both willing and able to assume the duties.

Flexner was averse to pondering contingencies. He was confident that Birkhoff could be secured. Moreover, offering the first position to Veblen made no sense. Such a move would fail to elevate mathematics in the city of Princeton and was likely to generate bad feelings from the university. Flexner's response was to deflect Veblen's suggestion while maintaining their close relations and biding his time until an offer could be made to Birkhoff.[51]

As the January 1932 Board meeting approached, Veblen was better informed than any trustee as to Flexner's thinking on the Institute. He was also more influential. Veblen's emergence as a confidante occurred as Flexner was losing his battle to include faculty on the Board. From the beginning Flexner had advocated a substantial voice for the faculty. As long as no one was hired, the issue had been moot. With the first appointments finally on the horizon, the matter was coming to a head. All fifteen Board slots were presently occupied. To include faculty required the retirement of current members or an amendment to the by-laws.[52]

Fortunately for Flexner, the by-laws had provided him with considerable freedom to operate as Director. Now he was pushing Maass and Hardin to draft an amendment that would enlarge the Board, making room for faculty. What he got, presumably at the direction of Bamberger, was a concession for up to three faculty to attend Board meetings in an advisory capacity, without the standing to vote. Flexner had lost a battle, but he continued to hold the confidence of the Board. His report from the previous meeting was adopted in principle, despite its advocacy of faculty trustees. [53]

More significantly, the trustees were endorsing the recommendation that the Institute begin its operations with Schools of Mathematics and Economics. Flexner requested and received permission to put forward, at the next meeting, the nomination of the first professor of mathematics. Regarding the other program, Flexner revealed that he would not be ready, at that time, to proceed on economics.

The decisions to take up mathematics first and economics second were entirely due to Flexner. He had spent a solid year of consultation and contemplation on the question, and it is difficult to imagine another non-mathematician reaching the same conclusion. There had been no ground swell of support for mathematics, except among mathematicians. When, in August, Flexner pitched the idea to a few trusted advisors, the reaction was mixed. Aydelotte and Frankfurter were supportive, though not particularly enthusiastic. Beard, who was advocating a program in civilization, urged Flexner to "chuck mathematics and take economics." No trustee possessed any knowledge or expertise in the subject.[54]

In selecting mathematics there were risks. Flexner himself could only direct in the most superficial of senses. So much was riding on Birkhoff. Even if he did accept, his was a name that was only known and respected by mathematicians. The Institute would begin with a first-class program whose accomplishments would not be tracked by the larger intellectual community.

While Flexner often boasted that his goal was to keep the Institute out of the press, he was being somewhat disingenuous. Substance and quality were his paramount values, but he was fully aware that his Institute was a production that was playing before an audience that included the founders, trustees, university presidents, and serious scholars. Flexner appreciated the role of publicity and was savvy in his devices for obtaining it.

This made economics all the more important. If Institute scholars solved the depression problem, it would indeed be a triumph on the magnitude of antibiotics. The parallel accomplishments of the mathematicians would then be recognized, if not comprehended. Flexner was looking for this one-two punch of mathematics and economics to place him with Gilman in intellectual history.

The odds for success in economics were much longer than in mathematics. There was no Birkhoff who, if procured, was virtually certain to lead a successful program. Moreover, if Birkhoff failed to crack some particular problem, he could be expected to succeed on another. Flexner was banking that he would identify and enlist the personnel who would then go on to solve the seminal economics problem of the time. After a full year of searching there remained more questions than answers. His thinking was that Viner was the ablest American, but was he up to the challenge? Perhaps in tandem with some European, but the notion of bringing in people with the charge to collaborate on a particular problem was anathema to Flexner's governing philosophy.[55]

Hermann Weyl in the 1930s. (Courtesy of Nina Weyl.)

CHAPTER FIVE

THE FIRST HIRES

The January 1932 trustees meeting provided Flexner the authority to negotiate with Birkhoff and to formulate a program in economics. Rather than advance these agenda, Flexner's next move was to board a train for the long ride to California. Over the previous decade, the California Institute of Technology had become an important center for research and graduate education in physics and astronomy. It was the only premier American university that Flexner had not visited, and he reasoned that his observations would complete his canvass of American institutions. He had contemplated the trip for some time, and, during the previous year, it would have fit in nicely with his visits to Harvard, Princeton, and Chicago. His current timing, however, was peculiar. The potential benefits of the extended journey appeared abstract and pale in comparison with the real opportunity to advance on Birkhoff's hiring. As it turned out the trip to California would have an enormous impact on the Institute for Advanced Study. It was there that Flexner met Albert Einstein.

Over the prior quarter century Einstein's theory of relativity had profoundly altered human understanding. His scientific achievements and personal mystique combined to penetrate popular culture and attain a notoriety that was unprecedented in history. No one in the world was so widely regarded as a genius. Flexner believed that the success of the Institute for Advanced Study hinged on his procuring a few geniuses. Einstein was the icon.

Albert Einstein was born March 14, 1879, in Ulm, Germany. His early life just barely reveals the characteristics that were to figure so prominently in his later identity. His parents were Jewish, but the household was not observant. As a student he indicated some talent in mathematics and science, but he was never regarded as a prodigy. Einstein's attitude and approach to learning seem to have conflicted with that of his teachers.[1]

When Einstein was 15, economic and business difficulties forced his parents to move to Italy, leaving Albert behind to complete his secondary education. Within a year he dropped out of school and joined his family in Milan. There he renounced his German citizenship, and lived with neither a country nor regrets. Einstein's multinational adolescence would continue in Switzerland, where he resumed his education. In 1900 he received an undergraduate degree from the Eidgenössische Technische Hochschule (ETH) in Zurich. His record was mediocre.

The twentieth century began with few prospects for the person who would later be acclaimed as its greatest scientist. Einstein was unemployed and without any savings. Complicating his predicament were his engagement to classmate Mileva Maric and his ambition to pursue independent research in theoretical physics. To increase his meager opportunities Einstein began the cumbersome process of obtaining Swiss citizenship. Meanwhile he eked out his own subsistence through tutoring and occasional teaching jobs.

While Einstein required few comforts for himself, his situation was becoming desperate. His fiancée Mileva returned to Serbia to have their first child. Meanwhile another classmate provided some hope. He offered to use his family connections to seek a position for Einstein at the Swiss patent office in Bern. It was not a great job, just a low-wage, entry-level position. Moreover, it would take more than a year to materialize, assuming it did. That this would remain Einstein's most attractive prospect gives some measure of his dire situation.

In the summer of 1902, Einstein began work at the patent office. For eight hours a day, six days a week, he and several coworkers examined patent applications as they arrived. Undoubtedly there were moments of humor as arcane contraptions were earnestly proposed to meet overstated needs. Then there were the rare, genuine discoveries. More typically, the duty was to decipher a description and determine the recommendation to provide to a superior.

Despite the low pay and what many would regard as tedium, Einstein found the duties agreeable. The stability of his position enabled him to marry Mileva and settle with her in Bern. It was not an environment for the incubation of world-class physics. Einstein had an undistinguished undergraduate degree, a 48-hour-a-week job, and the chaos of a young family. While highly regarded young scholars were guided in promising directions by professors at leading universities, Einstein relied on a couple of like-minded friends for intellectual stimulation.

It is truly remarkable that Einstein's research thrived under these circumstances. His personal and intellectual assets overcame the formidable disadvantages and distractions that he faced. While fulfilling his responsibilities to the patent office, Einstein pursued the development of his own original ideas. In the months surrounding his 26th birthday in 1905, Einstein submitted several papers for publication, including his work on the photoelectric effect and the special theory of relativity. The most immediate impact was the acceptance of another paper to fulfill the requirements for his PhD thesis in physics from the University of Zurich.

Appreciation of Einstein's discoveries developed slowly. In 1907 he was rejected for the postdoctoral position of *privatdozent* at Bern University. The following year he was approved for this nonremunerated title. Einstein was then eligible to give lectures and to receive nominal fees from attending students. Rather than return to academe, Einstein remained in his job at the patent office and continued his productive research at home.

In 1909 Einstein finally left the patent office to accept a faculty position at the University of Zurich. He rapidly ascended the academic ladder with a promotion to professor in Prague and then a return in 1912 to the ETH in Zurich. Zurich was a comfortable base for Einstein. While he normally worked without collaborators in his research, Einstein always enjoyed social and scientific interaction with selected companions. In Zurich he was reunited with a number of friends from his undergraduate days, whose company, to some degree, mitigated the discomfort he was experiencing from his crumbling marriage.

Meanwhile, Einstein's stature among scientists had grown enormously. In just a few years he had made the stunning advance from a hobbyist physicist to a regular on the list of nominees for the Nobel Prize. Along with this recognition came new employment opportunities. In 1913 the eminent physicist Max Planck offered an attractive package of positions in Berlin. It is difficult to reconstruct Einstein's analysis of Berlin versus Zurich. The improvements in compensation and status were likely of little importance to the unpretentious Einstein. More significant was the absence of teaching responsibilities and the intellectual stimulation of the Berlin scientific community. Thinking and talking about physics were Einstein's passions. For years he had stolen time for his research while doing less interesting work to support his family. Planck was offering ideal working conditions.

When Einstein moved to Berlin in 1914 at the age of 35, his life entered a new phase. He was established in an academic position worthy of a person with his discoveries. There were also accompanying changes in his personal life. Mileva remained in Zurich with the children, marking the formal separation from their marriage. In Berlin Einstein was more free to pursue his relationship with Elsa Löwenthall that would eventually lead to his second marriage.

Germany was the country of Einstein's birth and of his teenage expatriation. With his return, Einstein was forced to confront his ambivalence toward the culture and people. Departing the neutrality of Switzerland, he was quickly appalled by the enthusiasm he observed of a population ramping up for war. It was not just the chauvinism of people in the street. Planck and other intellectuals, such as the mathematician Felix Klein, went on record supporting the German aggression of World War I. Einstein reacted by adopting a lifelong commitment to pacifism. An apt metaphor for how he dealt with these issues was his decision to restore his German citizenship while maintaining dual Swiss standing.

Amidst the turmoil in the world and in his personal life, Einstein continued to work out his ideas on physics. In 1915 he completed his seminal work on general relativity. According to J. Robert Oppenheimer,

> In the general theory of relativity he created the single greatest theoretical synthesis in the whole of science, giving us a new understanding of the universality of gravitation and a new view of the cosmos itself. Unlike most great discoveries in science, Einstein's general theory could well have lain undiscovered but for his genius.[2]

Under the Einstein models of relativity, physics is more complicated than under the traditional principles of Isaac Newton. The differences, however, largely emerge in extreme situations, such as enormous mass or velocity. For example, when speeds are not close to that of the 186,000 miles per second of light, then the Newton-Einstein differences can be negligible. In the early twentieth century it was especially difficult to create experimental situations to ascertain whether the Newton or general relativity concept ruled. One possibility involved observing the path of light originating from a star, passing near the sun, and then to the earth. With the Einstein model, the massive gravitational force of the sun exerts a measurable deflection from the path forecast by Newtonian laws. The operational difficulty was that the overwhelming brightness of the sun obscures the view of the observer on earth. To perform the experiment required the assistance of a solar eclipse.

World War I and the infrequency of solar eclipses delayed the experiment for years. It was not until 1919 that the pictures and measurements were made by British scientific expeditions to Brazil and Principe Island off the coast of Spanish Guinea. The verdict was pronounced at a meeting of the Royal Society. The data supported Einstein over Newton and the press took its lead from the presiding president, Nobel laureate J. J. Thomson. Thomson heralded Einstein's theory as "one of the greatest achievements in the history of human thought," comparing the German scientist to the British hero Newton.[3] Articles in the *London* and *New York Times* brought Einstein to the attention of the English speaking public. Meanwhile the eclipse data removed the doubts of most of the remaining skeptics among the experts.

This press coverage, late in 1919, marked the beginning of Einstein's worldwide celebrity. A number of factors, in addition to merit, affected the spread of Einstein's notoriety. One was the popular fascination that such a fundamental discovery was only comprehensible to as few as a dozen people. Another more complex ingredient was his associations with pacifism, Judaism, and Zionism. Einstein never actively practiced Judaism. When, several years earlier, he applied for the professorship in Prague, he was required to state a religious affiliation. At first Einstein responded that he had none. When he was informed that his answer disqualified him for the position, Einstein made the revision to Jewish. This incident summarizes Einstein's earlier attitude toward his religious heritage. He did not deny his Jewish background, but he did not consider it to be part of his orientation.

Einstein's perspective began to change early in 1919 as he and other Jewish intellectuals were solicited to support the Zionist movement. Einstein considered the issues and gradually began to acknowledge his belief in Zionism and to embrace his own Jewish background. Meanwhile, Germany was experiencing a strong post-World War I anti-Semitic backlash, and many Jews were trying to assimilate. Now, almost simultaneously, Einstein became a famous person who was proud of his Jewish heritage, while living in a country where virulent anti-Semitism was on the loose. Two episodes in 1920 illustrate the collision of these factors. Early in the year Einstein was heckled and his lecture disrupted at the University of Berlin. In August a conference was organized to trash the theory of relativity. Such incidents would persist throughout the decade as Adolf Hitler and his National Socialist German Workers' Party (Nazi) movement sought influence and followers. Einstein and relativity became a lightning rod for German anti-Semitism.

The impetus for Einstein's first visit to the United States was Zionism. Chaim Weizmann, who would later become the president of Israel, persuaded Einstein to join him in 1921 on a fundraising trip. The beneficiary was the

Hebrew University that was to be established in Jerusalem. On their tour of New York and other American cities, crowds pursued Einstein as if he were an entertainment idol. The attraction of Einstein was a major factor in the success of the money-raising venture. Added to the itinerary were a number of scientific talks that brought Einstein into contact with American universities and scholars.

Einstein was the most prominent intellectual figure in the world, but remained without a Nobel Prize until 1922. Even then, the award was made for the photoelectric effect and bizarrely excluded relativity. Both the delay and the relativity stipulation persist as embarrassing episodes in the history of the Nobel Prize. Explanations involve some combination of conservatism, ignorance, and anti-Semitism. Another anomaly was that the date for the award was set retroactively for the prior year, when no physics prize had been given.

The Nobel Prize further enhanced his stature, and invitations poured in for consultations abroad. Through the mid-1920s Einstein traveled extensively, promoting Zionism and attempting to explain his scientific discoveries. There were trips to Japan, Palestine, and South America. These journeys were curtailed in 1928 by heart problems, leaving the 49-year-old Einstein bedridden for several months. It would be some time before he was able to travel, first in Germany and then through Europe.

For many years Caltech had attempted to lure Einstein for a visit. What distinguished the Pasadena institution among the many suitors were its own Nobel laureate Robert Millikan and the astronomical observations that were taking place at its Mount Wilson Observatory. After several abortive attempts, arrangements were made for Einstein to visit for the first two months of 1931, in what Caltech hoped to become an annual winter residence.

Einstein's second visit to the United States, and first in a decade, was a huge story. Anticipation was so great that Einstein was deluged by fan mail sent to Berlin in advance of his departure. Flexner saw an opportunity to introduce Einstein to his recently incorporated Institute for Advanced Study. He used his connections with Caltech authorities to move to the front of the queue. Flexner offered the Newark residence of Bamberger and Fuld as a quiet refuge for the Einsteins to recover from their transoceanic voyage prior to proceeding west. The invitation was declined.[4]

Einstein's initial visit to Caltech was a success, and the parties negotiated a return for the following winter. Meanwhile world events were exerting their influence. In 1929 the Germany economy was in shambles and unemployment was rampant. Hitler exploited the widespread deprivation to promote, as a solution, a mild form of his nationalistic agenda. His message gained traction and the Nazi party made stunning advances in the Fall 1930 Reichstag elections. At this point the Nazis had the second largest representation in the legislature, but were outside the ruling coalition.[5]

The growth of political fringe movements, such as the Nazis, is difficult to predict. Many intellectuals, including Einstein, did not yet take Hitler seriously, putting forth a variety of reasons for the doom of his Nazi party. Just after the election, Einstein made his first trip to Caltech. When his boat docked in New York, he was besieged by a mob of aggressive reporters seeking a story for their curious readers. Einstein's impromptu press conference, carried on the front page of the *New York Times*, included predictable questions such as "Could you define your theory of relativity in one sentence?" Another reporter asked for Einstein's opinion of Adolf Hitler. Einstein replied: "I do not enjoy Mr. Hitler's acquaintance. He is living on the empty stomach of Germany. As soon as economic conditions in Germany improve, he will cease to be important."[6] As many others, Einstein did not foresee the depth of the economic problems, nor their ramifications.

By his second trip to Caltech, Einstein had witnessed another year of economic catastrophe and the parasitic growth of the Nazi party. Every day he came into contact with talented Jewish scientists whose religion effectively excluded them from employment. While Einstein's own situation seemed secure, he was deeply disturbed to live amidst such injustice. At times, over the past decade, he toyed with the idea of finding a more hospitable environment, but Berlin had a lot to offer and inertia is a powerful force. However, a person of his humanistic sensitivities could only endure so much. Einstein's second visit afforded Caltech a marvelous opportunity to retain the great physicist on a permanent basis. Nevertheless, the California university was itself a victim of the worldwide economic crisis. It was challenged each year even to make ad hoc financial arrangements for the Einstein visits. When Flexner reached Pasadena in 1932, the circumstances could not have been more propitious. His Institute was in the unique position of possessing an abundance of resources and a need for personnel. At the same time, Einstein was disposed to consider new opportunities for pursuing his research.[7]

Flexner's autobiography contains a widely accepted description of his first encounter with Einstein. In this account Flexner depicts himself as a passive actor who was the beneficiary of fortuitous events:

> Professor Einstein happened to be a guest professor at that time, but he was
> so lionized that I purposely refrained from calling on him. On the day on

which I was planning to leave Pasadena, Professor Morgan telephoned me and said, "You haven't talked with Professor Einstein."

Despite Flexner's qualms over disturbing the famous scientist, Morgan insisted on setting up a meeting. Flexner and Einstein discussed the plans for the Institute for Advanced Study. Upon their parting Einstein proposed a follow-up discussion for the spring when their schedules would intersect at Oxford. Flexner reflected, "I had no idea that he would be interested in being connected with the institute."[8]

This last statement is explicitly contradicted by Flexner's contemporaneous letters to family members.[9] Flexner instantly perceived Einstein's interest in a position at the Institute. With Bamberger and Maass, Flexner began referring to the "counting of unhatched chickens" as code for the Einstein recruitment.[10] Flexner had his reasons for publicly downplaying the significance of his Pasadena interview. The Caltech authorities viewed Flexner's visit as an opportunity to work together to permanently bring Einstein to the United States. When Einstein subsequently committed to the Institute, Caltech appealed for joint custody. An annual winter residence in Pasadena conflicted with Flexner's notion of the full-time approach. In the end Flexner's triumph was accompanied by an unpleasant correspondence with his former host.[11]

The record is less revealing as to Flexner's motives in traveling to California. Did he design the Caltech site visit as a cover for a scheme to recruit Einstein? The circumstantial evidence is persuasive, and Flexner's stealth is easily explained. An Einstein venture was a long shot, perhaps even a bit impudent. Informing Bamberger, or anyone else, in advance was likely to raise undue expectations. It was best to remain discreet until there was some prospect of a pay-off.

Flexner aspired to orchestrate some spectacular intellectual breakthrough to establish his Institute. The appointment of Birkhoff was an unimpeachable step, but it would only generate excitement among mathematicians. The prospects for solving the depression problem remained hung up on finding the right personnel, and even then success, if ever realized, would be deferred through years of work. Flexner could not rely on the economics route. The hiring of Einstein was another approach. He was the most coveted intellectual appointment in the world. Flexner was aware of the anti-Semitic forces repelling Einstein from Germany and the economic constraints limiting Caltech. The overlap of Flexner's oddly-timed visit with the Einstein residence is better explained as a covert mission than as a coincidence. Flexner was hoping to test the waters.

In their discussion, Flexner previewed his plan to open the Institute with a School of Mathematics. Einstein was a theoretical physicist whose work relied heavily on mathematics. His reaction was of enormous importance to Flexner. When Einstein fully embraced the notion of starting with mathematics, even asserting its advantages over physics, Flexner was elated. Einstein's approbation was tantamount to divine approval.

As the conversation naturally delved into personnel, Einstein's highest recommendation was for Hermann Weyl of Göttingen. It was a name that Flexner had heard throughout Europe and the United States. Just one month earlier he had learned from Veblen that Weyl might be interested in moving back to the United States. The discussion of personnel gave Flexner ample opportunity to make overtures to Einstein about joining a staff with Weyl and Birkhoff. As the courtship began, Flexner measured Einstein's responses and gained confidence that his flirtations were welcome. When the interview ended, Flexner, despite his later denials, was certain that Einstein was enamored with the Institute and likely to accept a future offer.[12]

Flexner had learned from Gilman the necessity of being flexible and opportunistic. His search "to find the best men" began with few preconceptions. The spring meeting with Birkhoff in Paris had triggered the decision to go with mathematics. With his mind settled on the only permanent personnel he envisaged in mathematics, Flexner had then shifted his attention to economics. Now, nine months later, he revised the plan. Rather than have mathematics and Birkhoff as a solid foundation on which to build other programs, the School of Mathematics became the feature attraction. With Birkhoff (the leading American mathematician and one of the best in the world), Weyl (the pre-eminent mathematician in the world), and Einstein (the greatest scientist of the century) the Institute would open with the standing Flexner had fantasized from the beginning. Its mathematics research program would be acknowledged as the finest in the world, an unprecedented achievement for an American institution.

Today, it might seem odd to locate the physicist Einstein in a School of Mathematics, but there was less specialization in the early twentieth century. At Göttingen the theory of physics was established as a vital ingredient of the mathematics program. Both Birkhoff and Weyl had contributed to mathematical physics. Title and protocol were of little importance to Einstein. Whether or not Flexner understood the distinction between theoretical physics and mathematics, he clearly construed the mission for each school in a broad sense. There was no problem in Einstein cohabiting with Birkhoff and Weyl.

A more delicate issue was the imbalance of foreigners on the faculty. Although the founding ideals explicitly rejected discrimination, Flexner's search for professors had been targeted at Americans. He was already conflicted over the consideration of European economists, but there is no indication that the Birkhoff, Einstein, Weyl slate raised serious concerns. Einstein's standing placed him above any such profiling. The chronology of Weyl's emergence offers insight into Flexner's thinking and priorities.

For over a year Flexner had been searching for a scholar with Weyl's resumé, in his mid-forties and arguably the greatest scholar in his field, yet Weyl did not receive serious consideration at an early stage. When, in Europe, Flexner determined that Birkhoff could be secured, the exploration of mathematics was curtailed. Flexner abandoned his plan to stop in Göttingen, passing up the opportunity to meet Weyl and other mathematicians. It was later that Flexner began to consider Europeans in conjunction with his needs in economics. The December meeting with Veblen placed Weyl on the screen. It was, however, the perception of Einstein and Weyl as complementary pieces that finally brought Flexner to visualize the full potential for the School of Mathematics. In penciling in Weyl, Flexner was making another concession. Up to this point he had scrupulously followed the Gilman approach of interviewing each candidate. Flexner had never even met Weyl.

If Flexner had met Weyl, he would have been impressed. The span of his work included physics and philosophy, as well as mathematics. He was a product of the pre-World War I German educational system that Flexner so admired. Hermann Weyl was born on November 9, 1885, in the town of Elmshorn, near Hamburg. Weyl attributed his adolescent intellectual awakening to the chance discovery of a commentary on Immanuel Kant's *Critique of Pure Reason*. Immediately he began to contemplate the notions of time and space.[13]

The director of Weyl's high school was a cousin of the great mathematician David Hilbert. This connection destined Weyl for study at Göttingen. When he arrived in 1904, Weyl was an unsophisticated 18-year-old with a letter of introduction to Hilbert. Weyl's timing was especially fortunate as Göttingen was then at its zenith. Hilbert's brilliant friend Hermann Minkowski had recently joined the faculty. The stimulation that Hilbert and Minkowski derived from their mathematical discussions animated the entire campus. Together with Felix Klein, the Göttingen professors provided an extraordinary environment for the study of mathematics. Doctoral and postdoctoral scholars flocked to the town to soak up its vibrant mathematical culture. Minkowski's premature death in 1909 was a blow from which Göttingen and Hilbert could only partially recover. World War I then eroded the entire German educational system. Göttingen's standing remained intact for decades, but there was something special about the period when Weyl was a student.

Weyl fondly recalled the inspiration he received from Hilbert, shortly after arriving in Göttingen:

> In the fullness of my innocence and ignorance I made bold to take the course Hilbert had announced for that term, on the notion of number and the quadrature of the circle. Most of it went straight over my head. But the doors of a new world swung open to me, and I had not sat long at Hilbert's feet before the resolution formed itself in my young heart that I must by all means read and study whatever this man had written. And after the first year I went home with Hilbert's *Zahlbericht* under my arm, and during the summer vacation I worked my way through it—without any previous knowledge of elementary number theory or Galois theory. These were the happiest months of my life, whose shine, across years burdened with our common share of doubt and failure, still comforts my soul.[14]

Hilbert, renowned as one of the last great mathematicians of breadth, always worked serially. He would spend several years completely absorbed in one area before shifting his attention to another. When Weyl arrived in Göttingen, his mentor's attention was focused on integral equations, having already completed his phases on the theory of invariants, algebraic number theory, and the foundations of geometry. Naturally, Weyl's thesis work would be in the current area of integral equations. However, his approach to a Hilbert major gave him interest and fluency in the wide range of mathematics that had been studied by his advisor.

It would take some time for Weyl to gain status within the tight-knit Göttingen mathematics clique. Neither his manner nor his dialect impressed his peers, but mathematical accomplishment was the true currency in this environment. Weyl completed his PhD in 1908, remaining at Göttingen as a member of the Hilbert group. He continued his strong work on integral equations, becoming a *privatdozent* two years later. Hilbert himself was then moving toward physics, and Weyl participated in these seminars.

Some indication of Weyl's mathematical power was provided by a course he gave in the winter term of 1911-12. Rather than select integral equations or mathematical physics, Weyl chose the topic of Riemann surfaces, which had emerged as a vital tool in the work of other Göttingen mathematicians. Weyl's development, published in book form the following year, provided a modern, new perspective on the subject. Like his advisor Hilbert, Weyl had the capacity to divine the essence of a subject and create a profoundly original foundation for its exposition. Weyl's book attracted a great deal of attention and would be an important factor in his eventual call to a professorship at ETH.

Göttingen offered impressive resources for the third thread of Weyl's intellectual growth. Edmund Husserl was a professor of philosophy who was an important figure in the development of phenomenology. Husserl also had close ties to the mathematics department. This provided Weyl with the opportunity for a significant non-Hilbert influence. Weyl's groundings in mathematics, physics, and philosophy, and his power to balance them, would become unusual features of his writings.

Weyl moved to Zurich in 1913. Accompanying him was his bride Hella Joseph, who gave up her study of philosophy under Husserl. Weyl overlapped with Einstein at ETH for a year, but there was no significant interaction between them as Weyl expanded his program to include aspects of number theory and mechanics. His work in Zurich was soon interrupted by World War I. A German citizen, he at first received a medical deferment, but was eventually drafted into the army. Many years later Weyl recalled that after serving for one year,

> In 1916 I had been discharged from the German army and returned to my job in Switzerland. My mathematical mind was as blank as any veteran's and I did not know what to do. I began to study algebraic surfaces; but before I had gotten far, Einstein's memoir came into my hands and set me afire.[15]

The Einstein memoir was the work on general relativity.

Einstein's new theory of physics carried formidable mathematical prerequisites. Weyl picked up the necessary geometry and added relativity to his repertoire. Not only did he work at enhancing the mathematical foundation, but he became a major player in the search for a unified field theory linking gravitation and electromagnetism. Weyl's profile was broadened by his book *Space, Time, and Matter* that became a popular resource for scholars seeking to penetrate the theory of relativity.

Weyl now was publishing major papers on analysis, number theory, physics, and logic. Unlike Hilbert he seemed to work in parallel on these diverse topics. By the age of 35 he was regarded as one of the leading mathematicians in the world. Offers flowed in from other universities. Particularly tempting was an opportunity in the early 1920s to join Hilbert at Göttingen. A number of nonscholarly factors were involved in choosing between the Swiss and German venues. In contrast to Einstein, Weyl had warm feelings toward his German roots, but Zurich offered a beneficial environment for his asthma. Hyperinflation was one of the serious economic problems plaguing Germany at the time of the offer, and Weyl was concerned about the future welfare of his family. As the brilliant scholar pondered the issues he was racked by indecision, flip-flopping between the two universities. In the end he remained at Zurich. Still it was curious that a mind so comfortable in the deep recesses of mathematics, physics, and philosophy was overwhelmed in making a binary choice.

By the mid 1920s Weyl's work on relativity led him back to pure mathematics. In what is now regarded as his greatest work, Weyl developed the structure of representations for finite groups. This culminated in a joint paper with Fritz Peter that laid the groundwork for the mathematical field of harmonic analysis. As ever, Weyl did not confine his thinking to one area. Accompanying his abstract theory of group representations was its elegant application to quantum mechanics. Weyl's extraordinary breadth and depth firmly established him as the heir-apparent to his advisor David Hilbert.[16]

Weyl accumulated an impressive collection of job offers that included German universities as well as the two leading American programs at Harvard and Princeton. Princeton made its move after the challenge grant from the General Education Board enabled the funding of several research chairs. For Weyl, career choices continued to pose his most challenging problems. Rather than choose between Zurich and Princeton, he deferred the decision by agreeing to accept, temporarily, the Princeton chair in mathematical physics for 1928-1929. This led to another year of vacillation and, at times, there was optimism in Princeton that he would remain. At the end of the year Weyl returned to Zurich where he again faced his nemesis.[17]

The German mathematical establishment viewed Weyl as the only suitable successor to his mentor Hilbert, who was reaching mandatory retirement in 1930. Unlike the circumstances of the previous Göttingen offer, where he would have been regarded as Hilbert's understudy, Weyl was now given a clear path to take his place with the Göttingen legends of Gauss, Riemann, and Hilbert. Together with this new inducement were the factors that had paralyzed him nearly a decade earlier. In particular, the German economy was in a disastrous state. Weyl struggled with these issues and seemed incapable of making a decision. Finally he could not escape the feeling of obligation to carry on the tradition of Göttingen and Hilbert. After considerable delay Weyl informed Zurich that he was leaving.

For Weyl, such decisions, rather than freeing him from his demons, were followed by anguishing waves of regret. When he and his Jewish wife Hella did settle in Göttingen, they found the economic and political realities were worse than anticipated. Then Hitler and the Nazis had their gains in the 1930 elections. A feeling of reluctant martyrdom set in for Weyl. How could he have let his loyalty to Göttingen and Hilbert pull him into such a bad situation? Weyl toyed with reneging on his resignation from the ETH, but his successor had already been selected. Complicating the negative recriminations were the wonderful aspects of returning to Göttingen. In particular, there was the daily stimulation from the exceptional group of mathematical colleagues. It was under these circumstances that Weyl informed Veblen of his second thoughts over declining the Princeton opportunity.

Although Flexner could not have appreciated fully the complex issues plaguing Weyl, he revised his plan to provide for a tripling of personnel in the School of Mathematics. The elevation of mathematics to feature attraction and the associated budget ramifications called for another look at the School of Economics. Flexner still had reservations over the tentative scheme of relying on Viner and a European to be named later. Another Caltech interview inspired him to complete the reversal of the roles he had previously cast for mathematics and economics.

Flexner's social science consultant Charles Beard was at Caltech to deliver an address. Over the summer he and Flexner had engaged in a spirited correspondence concerning a variety of foundational issues. Flexner had a deep respect for Beard's intellect and cherished their give and take. He saw in Beard a kindred spirit whose courageous outspokenness had made him an academic outlaw. From time to time Flexner had considered Beard for the faculty, but he worried whether someone in his late fifties would have the energy and fresh ideas to lead the Institute's highest profile program. With the resumption of their conversation in Pasadena, Flexner became convinced that Beard was precisely the caliber of scholar that he had been seeking, certainly superior to Viner.[18]

A new plan crystalized to open with a roster of four faculty. By himself, Beard would establish an interdisciplinary program in history, economics, and political science. The School of Mathematics with Birkhoff, Einstein, and Weyl was the centerpiece. Of great satisfaction to Flexner was the certainty that he had finally selected the right personnel. Moreover, the timing was perfect. From Pasadena, Flexner was routed through Arizona to confer with the vacationing Bamberger and Fuld.

In February Flexner met with the founders. He was back in the familiar position of making a presentation to millionaires on their own turf. This time, however, there were significant differences. In the past Flexner had always been a catalyst. He promoted a concept and financial arrangement that, if approved, would then be developed and administered by other hands. Never before had he devoted so much effort to crafting the details nor had he been preparing to direct the implementation. The latter fact made it especially important that this be an intermediate step. He was not asking for any money, but was preparing the ground for the Institute to receive more of the Bamberger fortune. Guiding him was an understanding of how Harper's insensitivity to Rockefeller had jeopardized the University of Chicago.

Bamberger for his part had already made a substantial down payment and provided Flexner with free rein and a generous expense account to develop the programs. Always cautious and conservative, Bamberger still expected to exercise oversight and be given the opportunity to approve crucial decisions. Such a juncture had now been reached while the American economy was mired in a miserable state. Thousands of banks had failed in the previous year and unemployment had reached 20%. The severity of the depression and its global character had finally become evident. Although still comfortably a multimillionaire, Bamberger's wealth had been eroded, raising legitimate concerns about the future.[19]

Flexner set out to kindle Bamberger's enthusiasm and gain the approval to proceed. He could still drop the names of Rockefeller and Gilman, but this time Flexner had a trump card. While cautioning not to count any "unhatched chickens," he revealed that Albert Einstein might be obtained for the staff. To an uneducated self-made Jewish American, the name was magic. Bamberger lowered his formal facade and enthusiastically embraced the plan for the School of Mathematics.[20]

The final obstacle had been cleared to open negotiations with Birkhoff, Einstein, and Weyl. Bamberger's only reservation involved the fiscal implications. Allowing for faculty salaries comparable to Flexner's $20,000 meant that these commitments alone encumbered about one third of the uncertain annual return on endowment. Flexner moved to allay the budget concerns by tabling the lower priority program in the social sciences. The first offers would be made exclusively in the School of Mathematics.[21]

When Flexner reached New York he wasted no time in proceeding on several fronts. He immediately updated Maass and Leidesdorf on his meetings

in California and Arizona. Neither of these board members had the veto
power of Bamberger and Fuld, but they were the only other trustees to receive
a preview of the job offers. On that same day Flexner wrote to Birkhoff,
inviting him to New York for further discussion. No agenda was given, but
the purpose was well understood.[22]

Another letter went to Weyl in Germany, describing the Institute and
concluding:

> We have decided after very mature reflection to begin in the field of math-
> ematics and for that purpose to bring together a few men in the prime of
> life and of outstanding significance. Your name has been suggested to me
> repeatedly in this connection both in Europe and in America. If the mat-
> ter interests you, I should be very happy to make a trip to Europe for the
> purpose of conference with you.[23]

With an Einstein meeting already slated for May in England, Flexner could
easily add a Göttingen visit to his spring itinerary.

The next few months would be vital in determining whether Flexner realized his dream of a pure Gilman graduate university. He had done all he
could and soon the decisions would rest with others. Three acceptances assured a successful start. At the other extreme, three declinations meant disaster. There was no contingency plan. Despite the risks of pursuing such huge
catches, Flexner remained confident. He was convinced that the Institute offered a situation that was in the best interest of each candidate. Birkhoff was
the first to face the decision.

George David Birkhoff took the midnight train from Boston, arriving
in New York on Friday morning, February 19. The meeting with Flexner
involved no new substance, but was rather the obligatory penultimate step
in the protocol of recruitment. Both men reiterated their enthusiasm and
vision for the School of Mathematics. Even at this stage Flexner was unable
to disclose a definitive site, stating that the possibilities had been narrowed to
Princeton and Washington. While no offer was made, it was clear that one
would follow and that it would be welcome. Birkhoff returned to Boston later
that day, permitting him to meet his Saturday morning classes at Harvard.[24]

On Thursday Flexner sent his apologies to Birkhoff that illness was delaying his next communication. Flexner was sufficiently healthy to travel to
Princeton over the weekend and survey possible sights with Maass, Leidesdorf, and Edgar (nephew of Louis) Bamberger. The delay in making the
formal job offer was likely in the hope that he would then be able to stipulate

Princeton as the venue. This was not the case, but progress was made. Early the following week Flexner wrote to both Bamberger and Birkhoff.[25]

The letter to Bamberger began with a rosy description of the Birkhoff meeting and a pre-hatched chicken count: "I have made no promises, but I am perfectly sure that on the terms which we discussed he will accept." Flexner went on to claim the kudos for the Institute "that a man of the highest standing should be willing to give up a privileged position in Harvard University in order to cast in his lot with us." Next came a more delicate matter. Flexner described the Princeton excursion with the three other trustees and reported that everyone concluded "that the site is an ideal one for an institution of higher learning." After alluding to the consistency of this verdict with the advice solicited from the outside consultants, Flexner gingerly came to the point. Princeton "was the wisest choice that can be made. However, we shall have no meeting of the Committee on Site until you and Mrs. Fuld return, for we shall do nothing at all without your knowledge and concurrence."[26]

Flexner's letter to Birkhoff included a brief summary of the Institute's mission that could just as well have been written by Gilman. Then, subject to approval by the trustees, Flexner offered Birkhoff a professorship at a salary of $20,000 along with a generous pension provision. Among the prerogatives was the authority for Birkhoff to select his own students. One stipulation delineated the Flexner philosophy:

> In view of the fact that both salary and retiring allowance are placed upon a basis new in this country, it is agreed and understood that all members of the teaching staff will be on a full-time basis, that is, while they will have unhampered freedom of speech and action, they will undertake no activity for the sake of financial profit. This does not exclude acceptance of a Nobel Prize . . . .[27]

Full-time was not to be construed as 24-7-52. Flexner was programming an extended summer break, during which professors were free to travel and work as much or as little as they chose. What Flexner was precluding was mercenary teaching and textbook writing.

Flexner closed the offer to Birkhoff with a view of his plans. "As the resources increase, other schools will be added, but development will be slow. We shall not hope to cover the whole field of learning." As to Birkhoff's department, "I may add that it is in our minds to round out the personnel of the School of Mathematics by inviting one or two other distinguished mathematicians to participate in its development. It is impossible for me to say

at the present who they will be." Incredibly, Flexner was not confiding in
Birkhoff his intention to pursue Einstein and Weyl.

Birkhoff was excited about the opportunity promised by Flexner and the
Institute. He was inclined to accept the offer, but there was much to consider.
Harvard was the oldest and most prestigious university in the United States.
Its mathematics department was the strongest. Relinquishing such security
for a promising concept was a big step. There was also Boston. He and his
wife had lived there for 20 years and their daughter was about to announce her
engagement to a member of a prominent Boston family. Could they leave for
a project in which even the site remained unsettled? Along with these issues
Birkhoff weighed the $20,000 salary against the $12,000 ceiling at Harvard.

Birkhoff estimated that he needed one week to reach a decision. The first
step was to see what action Harvard might take to retain one of its best scholars. From Birkhoff's perspective, he had accepted a demotion and pay cut
to come to Harvard. Through hard work he had achieved considerable success. It was reasonable to expect some countermove, or at least blandishment.
Representing the Harvard administration were President A. Lawrence Lowell
and mathematics department chair Julian Coolidge. Birkhoff met first with
Lowell, on Friday March 4.[28]

Over his long tenure as president, Lowell had clashed many times with
Flexner. In particular, Flexner was critical of practices by the Harvard Medical
School. He admired Lowell's predecessor and missed few opportunities to
suggest that the university had drifted off course. When Birkhoff described
his marvelous opportunity, Lowell was unimpressed. No serious scholar could
abandon Harvard and its 300-year tradition for an upstart project headed by
a charlatan like Flexner. He saw no need to make any accommodation and
was confident that Birkhoff would come to his senses.[29]

Over the weekend Birkhoff went to Rhode Island to seek objectivity from
an old friend. R. G. D. Richardson was the dean of the Graduate School at
Brown and secretary of the American Mathematical Society. Richardson was
enthusiastic about the Institute. He felt that acceptance of Flexner's offer was
in the best interests of both Birkhoff and mathematics.[30]

On Monday, March 7, Birkhoff met with Coolidge and informed his
department chair of the offer. When the reaction was more of the *Harvard
attitude*, Birkhoff affirmed the earnestness of his deliberation. It would take
a few hours for this to sink in with the administration. At this point, however, Harvard was daring Birkhoff to leave. Later that day he took up the
challenge. Birkhoff wrote to Flexner, accepting the offer. He did not plan to

notify Harvard until he received formal approval from the Institute trustees, scheduled to meet the following month.[31]

There was an additional Monday meeting at Harvard. Unaware that Birkhoff was drafting his acceptance, Coolidge reviewed the situation with Lowell. They acknowledged the seriousness of the threat, but not the worthiness of the adversary. The result was a Tuesday morning letter from Coolidge to Birkhoff that was a stunning blend of arrogance and paternalism. Coolidge rehearsed "Aesop's Fable of the dog crossing a stream with a large piece of meat in his mouth which he dropped in order to grap at the shadow of meat which he saw in the water."[32] After touching on Birkhoff's ingratitude and trashing Flexner, Coolidge closed with the following paragraph:

> You will understand from all this that we view your proposal with nothing but dismay. I think we have shown in the past a sincere wish to help you attain all possible success. I think you can safely trust us to act in the same way in the future. It may be that Mr. Lowell did not immediately understand your view of this question. I think I can assure you, after talking with him, that he understands it thoroughly now. He is a man of very wide academic experience. He has faced many such situations in the past. Why is it not reasonable to believe that he sees more clearly than you do in the present situation? Let me beg you, therefore, to reflect very carefully on what you are doing, and take counsel of your mature judgement which weighs solid advantages rather than imagination which may be allured by glittering promises. I know that Mr. Lowell will be glad to talk further with you at this time. You will surely make no decision without consulting his wisdom again.

The ambiguity of Birkhoff's intentions emerged on Wednesday. Citing the letter from Coolidge, Birkhoff wrote Flexner requesting a suspension of his decision, pending an obligatory meeting with Lowell. Flexner's response to this second letter was as gracious as to the acceptance. He fully supported Birkhoff's reconsideration and expressed the desire that the Harvard authorities receive every courtesy. Despite this setback Flexner remained confident that Birkhoff would eventually cut the cord.[33]

The confidence lasted for one day. Birkhoff was parsing the offer and had a question. Harvard's sabbatical plan provided a year's leave at half pay. What could he expect from the Institute? Flexner had not yet formulated a policy, but he now realized that the battle had been joined. He immediately raised Harvard with a year at full pay.[34]

Birkhoff was struggling to reach closure. He obtained an appointment with Lowell for Wednesday, March 16. On Friday he and Richardson went to New York and conferred with Flexner. Birkhoff pondered the decision over the weekend and again decided in favor of the Institute. When Flexner received the second letter of acceptance on Tuesday, he had every reason to rejoice. It had been a tough struggle. Neither Hopkins nor Chicago had succeeded in luring away Harvard faculty. The Institute had one of their best. It was the most significant event since the decision, two years earlier, by Bamberger and Fuld to endow the project. Flexner reported the triumph to the vacationing founders at their hotel in Los Angeles.[35]

The celebration was short-lived. As Birkhoff explained to Flexner, everything changed on Thursday afternoon. "A very important item of fact concerning the situation here, affecting my whole attitude, was brought to my attention, and I went to New York to see you the same evening."[36] Birkhoff had decided to remain at Harvard. This time the decision was final.

It was not the first time that Birkhoff had handled a job offer in a questionable manner. Recall that the Princeton University administration was frustrated by the circumstances under which he arrived and departed from their institution. In both of those cases Birkhoff's actions could be justified or explained by miscommunication. However, to twice retract an acceptance was a blatant exhibition of bad faith. The following interpretation of Birkhoff's behavior is built upon his cryptic words to Flexner.

Birkhoff's first reversal was in reaction to the letter from Coolidge. Coolidge argued that propriety dictated a further meeting with Lowell. If there were one consistency in Birkhoff's career dealings, it was that his actions were driven by factors other than courtesy. As meager and condescending as the overture provided by Coolidge, it was the first indication that Harvard might consider some accommodation. For Birkhoff this was worth exploring. Whatever, if anything, was offered proved unsatisfactory.

Birkhoff's final reversal came in response to some improvement in his situation at Harvard. The nature of the "item of fact" is unclear. It appears to have originated on the Harvard end, rather than in response to a non-negotiable demand. Birkhoff would later inform Veblen that Harvard had made no financial accommodation. Nevertheless, something did occur that changed Birkhoff's "whole attitude." Speculation as to the source begins by noting possible professional incentives that may have arisen at this time.[37]

Perhaps it was a coincidence, but one year later Harvard conferred two of its most prestigious honors on Birkhoff. At the 1933 commencement he was awarded an honorary degree. At the same time he ascended to the chair of Perkins Professor of Mathematics. The prospect of either recognition could fit Birkhoff's explanation to Flexner. Another possible link involves Birkhoff's son Garrett.[38]

Garrett Birkhoff was a talented young man who completed his undergraduate work in mathematics at Harvard in 1932. Garrett was interested in continuing his study at Cambridge University. One avenue for doing this was the recently endowed Julia Henry Fellowship. Only one Harvard student was to receive the award. At about the time that his father was negotiating with Lowell, Garrett was interviewing with the president for the fellowship. He was successful. When he returned to the states the following year, the timing coincided with Lowell's creation of the Society of Fellows. Garrett's selection as one of its first members supported three years of independent study. Upon the completion of his term as a Fellow, Garrett joined the Harvard mathematics faculty. Even without a quid pro quo in 1932, the father may have foreseen parts of his son's trajectory and wished to be at Harvard as it unfolded.[39]

One other interpretation of Birkhoff's actions must be considered, even if it is inconsistent with the explanation given to Flexner. Among the mathematics community, there was a widespread belief that Birkhoff held anti-Semitic feelings. It is possible that Birkhoff was reluctant to join an enterprise funded and directed by Jews. His understanding of the Jewish involvement may have increased over time. At the beginning of the negotiations, Flexner did not disclose Einstein's possible participation. The closely held secret could have been divulged in the final push to attract Birkhoff.

Whatever the motivation behind G. D. Birkhoff's decision to remain at Harvard, the victim of his method was Flexner. Birkhoff was the only ideal candidate whom Flexner had identified for his faculty. He was an American of the right age with both the scholarly and mentoring credentials. Although Einstein and Weyl had greater scientific accomplishment, they were foreign and less successful as thesis supervisors. Age was an additional issue with Einstein and Beard. It was no coincidence that Birkhoff received the first offer. He was the most fundamental ingredient of the School of Mathematics, which itself was, at this point, the entire Institute for Advanced Study.

Although Flexner maintained a brave face, the Birkhoff reversal was a devastating blow. It fell to Abe's pathologist brother Simon to console and perform the postmortem. Simon recognized the damage from Birkhoff's action. He wondered whether Abe's controversial role as a critic interfered with his recruiting efforts. Simon was diplomatically suggesting that animus lay behind the Harvard campaign to retain Birkhoff. Simon advised that Abe lower his profile as a university reformer.[40]

In his response Abe admitted that "despite an air of self-confidence, without which I should never have been able to do anything, I have all my life been haunted by the fact that I have never been on the inside of any going institution." He went on to acknowledge and defend his past activity.

> Of course, I know how university presidents feel about me. Those with whom I have found it possible to cooperate at the General Education Board think I am a little Jesus Christ. Lowell would characterize me very differently, but this is not so much because of my book as because of the fact that for many years we refused to give money to the Harvard Medical School. I was opposed, and he knew it, and time has proved me right . . . .[41]

Certainly Lowell and Flexner's other Harvard detractors had worked on Birkhoff. In any event, Abe reasoned that if Birkhoff were so easily alienated, then perhaps the Institute would be better off without him.

There remained considerable question as to where the School of Mathematics stood without Birkhoff. One of Flexner's fundamental axioms was "no subject will be chosen or continued unless the right man or men can be found." These were the words Flexner wrote for the trustees, prefacing his selection of mathematics as the first program. At the time of the drafting, six months earlier, Birkhoff was "the right man." Since that time Einstein and Weyl had entered the picture, but neither could replace Birkhoff. Without a contingency program to substitute for mathematics, Flexner downplayed the loss of Birkhoff and resumed his pursuit of Einstein and Weyl.

In his unsuccessful recruitment of Birkhoff, Flexner had worked entirely on his own. Flexner informed Oswald Veblen of his intentions, but stressed the confidentiality of the operation. While Veblen was unable to resist hinting to his old friend Birkhoff that an offer was forthcoming, he refrained from any lobbying. The approach to Weyl was different from the start.[42]

It was Veblen who had volunteered his services to Weyl to act as his agent in America. Flexner entered the picture as Weyl began a sequence of reactions to Veblen. It was quintessential Weyl. In January Weyl responded that he had made a mistake in leaving Zurich for Göttingen. There were the economic problems and then there was the possibility that Hitler would gain control. With his Aryan-Jewish marriage the future looked bleak for his family. As to the possibility of returning to America, he was leery of New York, Yale, and Chicago. Princeton, however, was more appealing. Weyl wrote that he was now agreeable to the Princeton terms he had rejected a few years earlier.[43]

One month later Weyl had a characteristic change of heart. It appeared to him that the influence of Hitler and the Nazis had peaked:

> I am a little bit more at ease concerning the political future of Germany. I now have the impression that Hitler is lacking the courage to take responsibility for carrying out his radical ideas and that he is happy if he can stay in the opposition. If National Socialism will be established then probably only in a very much watered down form which will be at most economically dangerous.[44]

Weyl then discussed visiting America on a purely temporary basis. This February letter crossed in the mail with Flexner's overture for joining the Institute. When Weyl replied in March to Flexner's letter, he had shifted to a position between the extremes of the previous two months. Subject to obtaining more details about the Institute's location, library, and other faculty, Weyl proposed an arrangement more suited to his indecision. He might take leave from Göttingen and spend a year in residence at the Institute. During this period he would make up his mind whether to remain or return to Göttingen.[45]

Given Weyl's difficulty in coping with these matters and the unstable political situation in Germany, it was impossible even to extrapolate his feelings to April. By June, when Flexner planned to visit him in Göttingen, the Institute might no longer be part of his deliberations. Under these circumstances, and in the midst of the Birkhoff negotiations, Flexner was drawn to Veblen for support. Veblen was scheduled to reach Göttingen in April. He and Flexner agreed that Veblen would lead their dual recruitment of Weyl.[46]

Over the winter of 1932 Flexner became increasingly dependent on Veblen. Their relationship had arisen out of a mutual interest in the promotion of American research. As Princeton and mathematics moved from hypothetical to real components of the Institute, Veblen graduated from consultant to confidante. Flexner began to rely on Veblen as a sounding board. When recruiting in mathematics offered unexpected challenges, Flexner took the unprecedented step of permitting Veblen to act on his behalf with Weyl. Veblen's role with the Institute would continue to expand.

Oswald Veblen was of Norwegian descent. He was the nephew of the well-known economist Thorstein Veblen. Oswald was born on June 24, 1880, in Decorah, Iowa. At that time his father Andrew was teaching mathematics and English at Luther College. Shortly after Oswald's birth the family moved to the east while Andrew pursued graduate work at Johns Hopkins.

Oswald Veblen in the early 1930s. (From the Archives of the Institute for Advanced Study.)

They returned to Iowa in 1883 when Andrew joined the physics faculty of the University of Iowa. Oswald's youth was spent in Iowa City. He obtained a BA from the University of Iowa and then another BA from Harvard.[47]

In 1900 Veblen began graduate study in mathematics at the University of Chicago. He arrived prior to Birkhoff and also worked under E. H. Moore. Moore was the leader of American mathematics at the time. His interests reflected mathematical developments taking place in Europe. One of these areas was the foundations of geometry, where there were efforts to identify an ideal collection of axioms for Euclidean geometry. Veblen's thesis established independence of a system that featured an economy in the number of axioms and undefined terms.

After receiving his PhD in 1903, Veblen remained on at Chicago for two years. Looking back at this period, there were already indications of his prowess as a mentor and scholar. Veblen supervised the thesis of R. L. Moore, who was to become a leading American mathematician. Among Veblen's mathematical accomplishments was the first rigorous proof of the Jordan Curve Theorem. This intuitively obvious, but deceptively deep, result asserts that every simple closed curve separates the plane into two connected regions.[48]

In 1905 Veblen was recruited by H. B. Fine for one of the four new preceptorships in the Princeton mathematics department. The salary was \$1,500 with weekly duties involving six hours of class room time and nine hours of individual tutorial work with students. Fine's appointments clearly mark the beginning of Princeton's rise to prominence in mathematics. Suddenly a department that had produced little research was home to the most promising young mathematicians in the country. Within a few years other opportunities arose and two of the preceptors departed. In both instances Fine succeeded in obtaining strong replacements, including Birkhoff. When Yale approached Veblen, Fine countered with a promotion to professor in 1910.[49]

During Veblen's first five years at Princeton, the mathematics department had undergone a striking makeover. It became one of the leading centers for American research, a station it has not relinquished. Veblen took advantage of other campus opportunities as well. He met his future wife Elizabeth Richardson when she visited her physicist brother from her home in England. Veblen's entire outlook was international.

In 1913 the Veblens traveled through Europe. Oswald's letters to Birkhoff include impressions of European mathematicians and their regional culture. In Scandinavia Veblen was struck by the grace of Mittag-Leffler and the formality of the mathematicians at a conference. Everyone wore frock coats. Papers were presented in their entirety without follow-up discussion. In contrast, speakers at Göttingen were constantly interrupted in a manner that reminded Veblen of his time with Birkhoff at Princeton. In Berlin he found an atmosphere with less pressure than Göttingen.[50]

Veblen was quite impressed by Hilbert, noting that his manner evoked that of E. H. Moore. He was disappointed to discover that the other Göttingen mathematicians only explored the areas in which Hilbert had worked. In particular, they acknowledged the importance of Birkhoff's recent proof of Poincaré's Theorem, but showed little understanding of his papers. Veblen believed that American mathematicians were competitive with those at Göttingen, excepting Hilbert.

At the time of his European tour, Veblen's best known work was in projective geometry. He had published a book on this subject in collaboration with another of the original preceptors. Klein was the only member of the Göttingen faculty who appeared to have any familiarity with the book, and his understanding was superficial. Veblen came away with admiration for the community involvement in mathematical research at the German institutions, but he deplored the insularity of the Hilbert major. Over time he mellowed on the latter point, and offered Göttingen as a model for American universities to emulate.

Veblen himself was in the midst of pursuing a new research direction. During Birkhoff's time in Princeton they had discussed the work of Poincaré.

Poincaré had discovered a powerful interdisciplinary approach to mathematics that was known as analysis situs. The techniques were difficult and inaccessible. One of Veblen's gifts was a knack for recognizing merit in people and ideas. Some of his greatest accomplishments involved his role in the realization of potential. With analysis situs (now known as algebraic topology) Veblen worked to establish a foundation as well as obtain new results. Joining him at an early stage was a brilliant young Princeton graduate student. James Alexander did his thesis in a different area, but began collaborating with Veblen on analysis situs. Alexander would soon move out on his own and obtain some of the field's most fundamental discoveries.

When Veblen was invited to give the prestigious Colloquium Lectures to the American Mathematical Society in 1916, he chose the topic of analysis situs. These lectures and his subsequent monograph became the first primer on the subject. Through them, an entire generation of mathematicians was given the opportunity to pursue the techniques originally conceived by Poincaré. Many availed themselves and Veblen became one of the most influential mathematicians in the United States.

The tall, cosmopolitan Veblen exerted a forceful presence. Not only was he a scholar and mentor, but Veblen was an able administrator as well. It was typical for someone with these skills to become a department chair and gradually be diverted into management of the university. At Princeton, however, Fine and L. P. Eisenhart more than covered the mathematics department's share of duties in administration. While Veblen neither served as a department head nor dean, he played an enormous role in identifying Princeton's mathematics personnel.

One of the primary beneficiaries of Veblen's administrative skills was the American Mathematical Society. He began a two-year term as president in 1923. Veblen recognized that the society was in financial trouble. The $6 membership fee was inadequate to meet the increasing expenses for journal publication and other activities. Rather than increase the dues, Veblen led an endowment drive that placed the society on sound financial ground.[51]

Veblen's other agenda was to upgrade the environment for mathematical research in the United States. The situation at Princeton illustrated the obstacles. Only Fine, Eisenhart, and Veblen had offices on campus. Others did their research at home. Junior faculty faced teaching loads of nine hours per week, largely involving service courses to freshmen. Although Princeton was ranked second to Harvard among American mathematics departments, all of the research was performed by just a half dozen faculty members. At other universities the research faculty was smaller and teaching loads were higher.[52]

Veblen was especially sensitive to the plight of young scholars. It was through his initiative that the National Research Council began to include mathematics among the subjects for postdoctoral fellowships. This was an important step, but Veblen was seeking some mechanism for creating a mathematical research community such as he had witnessed at Göttingen. His next move was the proposal for an institute at Princeton devoted to research and graduate education in geometry and analysis situs. The faculty was to consist of himself, Alexander, Lefschetz and other leading scholars, both junior and senior. The proposal failed to receive General Education Board support.

Further progress was made with the creation of the research professorships and the construction of Fine Hall. As the Fine Professor, Veblen had no formal teaching duties. He typically gave one advanced lecture a week and supervised graduate students. Fine Hall provided offices for all of the research faculty. The Common Room, afternoon teas, and Mathematics Library offered an environment in which a mathematical research culture spontaneously generated and thrived. It was much of what Veblen hoped to achieve with his institute. All that remained was to release the other faculty from their teaching duties and to add personnel. When Fine Hall opened in 1931, there was little prospect of obtaining these concessions from the university. Then Flexner appeared with an endowment and a disposition toward supporting mathematical research in Princeton. Veblen saw an opportunity to achieve his institute through Flexner's Institute.

Veblen and Flexner held the same philosophy, but their notions on personnel revealed significant disparities. Flexner's criteria (best in the country or world) were met by Birkhoff, Einstein, and Weyl. The Princeton mathematicians were just below the threshold. Moreover, Flexner was averse to poaching on his putative hosts. He envisaged his elite trio and the larger Princeton group collaborating as colleagues in Fine Hall, but having their affiliations clearly separate and understood. The loss of Birkhoff compelled him to revisit the plan.

When the Institute Trustees met in April 1932, Flexner reaffirmed his commitment to mathematics and economics. He regretted that he was not yet prepared to nominate the first mathematician. The Birkhoff negotiations were never mentioned. Flexner then declared his intention to go abroad for further consultations. In other business, the Committee on Site inched closer to committing to the Princeton location. Princeton was acknowledged as the best prospect, but additional investigation was required. The trustees authorized the Executive Committee to act on a forthcoming recommendation of the Committee on Site.[53]

It was an obvious disappointment that Flexner had failed to deliver his first mathematics appointment. Bamberger bought some time by disclosing that the Director was on the verge of action. This also raised expectations for the trip to Europe. For Flexner to maintain the confidence of the Board, it was vital that he reach an agreement with Einstein or Weyl.[54]

As earnestly as Flexner was attempting to attract these men to the Institute, a more powerful force was pushing them away from Germany. There had been several political developments since Weyl's ambivalent March letter to Flexner. One week later came the German election for president. Hitler received a solid 30% of the vote, placing second to the 84-year-old incumbent Paul von Hindenburg. Since Hindenburg just missed a majority, a three candidate run-off was scheduled for April 10. Hindenburg won, but Hitler's charismatic campaign increased his share to 37%, representing over 13 million voters. In multiparty Germany, Hitler and the Nazis had attained critical mass.

Despite these events, Weyl remained inscrutable. On April 24 Veblen reported to Flexner from Göttingen: "Have talked somewhat with Weyl. I think he would not be averse to the move but his wife is very well satisfied with her position here. They don't seem to be as much alarmed over the political situation as I expected, in spite of the fact that Hitler is expected to win in the elections today."[55] The elections, which were local, reinforced Hitler's showing earlier in the month. The Nazis won a plurality in Prussia.

Flexner followed these events, but could not be certain how they affected his other recruit. Einstein had returned from California just in time to witness Hitler's strong showing in the run-off. Soon after his arrival, Einstein was off to England for a residence at Oxford. There he met, in May, with Flexner.

For Flexner it was a crucial meeting. He needed Einstein, but had yet to make an outright offer. They discussed Weyl and other Institute matters as Flexner dropped his hints and sought an opening. Finally he summoned the courage to offer Einstein a position, accompanied by a blank check as to salary. Einstein was interested, but requested that his decision be postponed until after his return to Germany the following month. As anxious as Flexner was for an answer, the delay was a reasonable request. He could easily drop in on Einstein after seeing Weyl in Göttingen.[56]

During his two weeks in England, Flexner discussed mathematics and economics with a number of people. To the mathematicians, Flexner floated a new name. Even with Einstein and Weyl, the mathematics faculty could not be complete. The Institute was to be an American institution and Flexner was convinced that the faculty required American representation. In England and then in Paris, Flexner sounded out opinions on Veblen.

Another matter on Flexner's mind was the deteriorating United States economy. While in Europe he did his best to keep track of the markets back home. All of the news was terrible, and indicated that the Institute assets were in decline. With Einstein holding a blank check it was difficult to reckon either side of the budget ledger. The only certainty was that the Institute was still without any faculty. When Flexner arrived in Göttingen at the end of May, he was under considerable pressure to hire a mathematician.

In Göttingen, Flexner saw Veblen and Weyl. It was his first meeting with Weyl. They became acquainted and discussed the plans for the Institute. Flexner portrayed a small mathematics faculty concentrated in geometry. Geometry was but one of Weyl's many interests. He expressed a desire to include an algebraist among the faculty. Another ingredient he stressed was youth. This made the 34-year-old algebraist Emil Artin of Hamburg an ideal choice for Weyl. Flexner took the recommendation under advisement.[57]

Weyl was interested and conflicted. He described his needs to an empathetic Flexner. Salary was important. German inflation had wiped out his savings. Weyl wanted a comfortable, secure living for his family, and the means to educate his children. It was almost a verse from Flexner's book on universities, a book that Weyl read prior to the visit.[58]

As always, Weyl was reluctant to make a commitment. Now he proposed a three-year trial period, during which he retained his options for both the Institute and Göttingen. Any such arrangement required approval from his colleagues and the Prussian Education Ministry. Flexner was well acquainted with the education ministry, knowing the process and the players. It would take some time to accommodate Weyl's issues. Flexner outlined the terms of a hypothetical offer. He would return later in the summer to continue the discussion.[59]

Another Göttingen negotiation was more clear cut. For some time Veblen had been keen to join the Institute faculty. His Princeton affiliation and standing behind Birkhoff had disqualified him from Flexner's earlier consideration. When Birkhoff withdrew, Flexner was forced to order his priorities. Veblen was the best available American, and he was respected by the European mathematical community. Moreover, he had been a valuable consultant and was much better suited than Einstein and Weyl for handling the administration of the mathematics program. The remaining obstruction was the Princeton authorities. Unlike his attitude toward Harvard, Flexner felt constrained to obtain Princeton's release of their leading mathematician. Flexner discussed the contingency with Veblen, who assured him that the Princeton administration would be amenable.

When the discussion reached salary, Flexner proposed $15,000. No doubt the bleak economic conditions figured into Flexner's adjustment from the $20,000 offered to Birkhoff. Flexner also believed that salaries should depend on individual circumstances and needs. As the Veblens were without children, he could justify a lower figure. Apparently these factors outweighed his symbolic goal of elevating professorial salaries to the level of institutional chief executives. In any event, Veblen was thrilled with the prospect of a 50% increase from his current $10,000. It is unclear what salary was mentioned to Weyl.[60]

When Flexner reached his next stop in Hamburg, he learned that a close relative had just died in Kentucky. The likelihood of an early return to the States injected a new urgency into his affairs. Flexner then completed two personnel matters. After speaking with some old friends at the University of Hamburg, Flexner concluded that Artin was unlikely to accept an offer. Flexner scratched Artin off his list. The other move was to follow through with a letter to Veblen, formally proposing the terms they had discussed. The offer was made subject to approval by Eisenhart, the Jones family (donors for Fine Hall and the Fine Professorship), and the Institute trustees. Veblen quickly replied with his acceptance, along with some talking points for persuading Eisenhart.[61]

With his wife in Vienna, mourning her uncle's death, Flexner abruptly left for Berlin. There, en route to comforting his wife, Flexner could make his rendezvous with Einstein. Just a few days earlier there was a significant political development. Hindenburg forced the resignation of the German Chancellor, replacing him with Franz von Papen. The installation of von Papen was widely understood as increasing the role of the military. While many viewed this as adding stability to the government, Einstein correctly foresaw the move as a prelude to a Nazi takeover. His future prospects in Germany had dimmed further.[62]

On Saturday, June 4, Flexner traveled to Einstein's country home in Caputh. It was raining and unseasonably cold. Flexner in his overcoat was greeted by Einstein in summer attire. To Flexner's concern that he might be chilly, Einstein responded that he dressed according to season rather than weather.[63]

This was Flexner's third meeting with Einstein, and it was likely his last opportunity for one on one persuasion. Flexner was desperate to reach an agreement. Einstein's blank check went beyond salary. Virtually any benefit or inducement was his for the asking. What Einstein did choose to request would require an enormous concession from Flexner. Einstein wanted an Institute appointment for his 45-year-old collaborator Walther Mayer. Einstein blamed anti-Semitism for Mayer's inability to obtain an academic position.

Flexner faced a dilemma. Nothing was more sacrosanct than the standards for his faculty. Mayer, who appeared to have the role of Einstein's assistant, did not meet the criteria. Yet Mayer was the price for obtaining Einstein. Flexner weighed the Einstein-Mayer package. Pragmatism won out over idealism and principle. He would have to find a place for Mayer in the Institute. It is unclear whether, amidst the language barrier, there was any specific understanding on Mayer's job title.

Einstein's salary was an easier matter. Flexner would gladly have supported a figure beyond his own $20,000. When he asked Einstein to name his price, the response was $3,000. Anticipating no difficulty in agreeing to a satisfactorily higher level, Flexner moved on to more controversial issues. A big problem was Einstein's other commitments.

Einstein's base was in Berlin where he was affiliated with several institutions. Additional obligations included the following winter at Caltech and one month annually at Oxford. There was ambiguity as to whether the residence in California was understood to continue in subsequent years. Flexner appreciated the sensitivity of Einstein withdrawing from prior commitments, but an Institute professorship entailed certain obligations. The most fundamental was to reside in Princeton for the academic year that extended from the beginning of October to mid-April. There was no difficulty with occasional trips or summering in Caputh. It is unclear what Flexner said or implied to Einstein about severing his other institutional commitments. Since Flexner was downplaying the obstacles, he probably abridged his sermon on the full-time approach. What Einstein understood was his responsibility to remain in Princeton for about one half of each year.

Flexner and Einstein had reached general agreement on terms. It was left to Einstein to deal with his other obligations, and to Flexner to draft the provisions. They parted with Einstein expressing his enthusiasm as "Feuer und Flamme dafür" (literally: fire and flame for it).[64] On Monday Flexner followed through with a letter. The pension arrangement was similar to that for Veblen, and the salary was $10,000. Although the latter figure was much higher than Einstein's request, it was surprisingly lower than what was agreed to with Veblen. Mayer was to receive $4,000 per year as Einstein's assistant.[65]

Delighted with his Berlin triumph, Flexner left for Vienna. There he was reunited with his wife, Anne. The couple spent considerable time apart, necessitated, to some degree, by Anne's playwriting career. Abe decided that he would return to handle matters in America. His last stopover was in Paris.

When Flexner arrived in Paris, he received the mail that he was anxiously awaiting. Subject to a few modifications, Einstein accepted the terms that were offered the previous week. The most notable issue involved Mayer. Einstein asked that Mayer's appointment be stipulated as independent of his own. On the amusing side was Einstein's objection that his pension allowance was too high. His other concerns were the starting date and his desire for protection from dealing with individual applicants.[66]

On June 14 Flexner acknowledged his acceptance of Einstein's amendments. As to the pension, Flexner pointed to the uncertainty of the economy. He suggested deferring the pension question and promised to reach a mutually satisfactory solution. Flexner also agreed to the independent appointment for Mayer. Flexner was conceding a permanent position for Mayer at the Institute. While title and standing were not addressed, it was reasonable for Einstein to infer that Flexner was agreeing to a professorship. These matters would soon become a source of contention as Flexner groped for a path that would satisfy Einstein, but exclude Mayer from the faculty.[67]

Despite his compromise over Mayer, Flexner was exuberant. The European trip had been a gamble. It was his last opportunity to deliver a big name to the October Board meeting, and there was no name bigger than Einstein. The Institute and its Director had established their credibility. Consider the scenario if Einstein had declined. What were Bamberger and the trustees to think of a one and a half year search to identify and steal a Princeton faculty member? As important as Veblen would be in charting the direction of the Institute, his appointment did not stand on its own. Together with Einstein it was perfect.

Flexner remained optimistic about Weyl. In their last communication he had invited Weyl to request a written expression of the Institute offer. Veblen's letters from Göttingen indicated that the Weyls were on the verge of committing to the Institute, even foregoing the three-year trial period. Flexner was still waiting for Weyl to request the terms.[68]

That was the situation when Flexner sailed for the States. Everything was top secret in order that Einstein have time to rearrange his other commitments. Flexner did confide in his immediate family. He also cabled his secretary to inform Bamberger and Fuld of the Einstein and Veblen developments. Flexner arrived in New York on Tuesday, June 21, and wasted no time in consolidating his gains. The following day he was in Princeton to see

Luther Pfahler Eisenhart, 1932. (Courtesy of the American Mathematical Society.)

Eisenhart prior to a Thursday luncheon scheduled with the founders, Maass, and Leidesdorf.[69]

Flexner attached great importance to his meeting with Eisenhart. He wanted Veblen for the Institute, but the long-term relationship with Princeton was more important. It was essential that the case be presented in such a way that Eisenhart give up Veblen freely, with no feeling of coercion or violation. If this could not be achieved, then the Institute would proceed without Veblen.

Flexner laid out his vision to Eisenhart. He described the beginning of an historic relationship. Einstein, Veblen, and (likely) Weyl would join the Princeton mathematicians to form the strongest group in the world. After Weyl completed the mathematics faculty, attention would shift to collaboration in other subjects.

The atmosphere was warm and friendly. Eisenhart was amused to learn that Weyl was "undergoing one of his periods of indecision."[70] As for Veblen, they had been colleagues since 1905 when both served as preceptors. Together they had followed Fine in building the department. After a quarter century of working together, Eisenhart could not stand in the way of Veblen's advancement. It was Eisenhart's magnanimity, however, that so greatly moved Flexner. Eisenhart was genuinely happy for Veblen's good fortune and he graciously offered space for the three mathematicians in Fine Hall.[71]

The key trustees must have welcomed Flexner to lunch as a conquering hero. Almost everything appeared set. Veblen was scheduled to go on the payroll in the fall. The official opening would be the following year when Einstein arrived. With space in Fine Hall available, Princeton was finally approved as the Institute's location. The only remaining detail was to wait for Weyl's acceptance.

Bamberger was ecstatic. To become Einstein's patron was an honor. It was not a circumstance for obtaining a bargain. He insisted that Einstein's compensation be raised to the same level as Veblen's. Once again Flexner succeeded in persuading Einstein to accept an increase.[72]

Meanwhile Einstein was notifying the authorities in Berlin and at Caltech. The German officials were content to retain a piece of Einstein that included summers in Caputh. The reaction in California was decidedly different. Millikan was disturbed that Einstein was terminating his winter Pasadena arrangement in order to accept Flexner's offer. The Caltech authorities initiated a campaign of pressure on Flexner.[73]

Millikan first pushed for Einstein to split his American time between Pasadena and Princeton. There was no way that Flexner was going to agree to such a settlement. The opening of the Institute was still one year in the future and the faculty roster consisted of just two names. Flexner was aiming to establish the Institute as a new intellectual center. To accomplish this he needed all of Einstein.[74]

Caltech then adopted a fall-back position of asking for Einstein in alternate winters. In the exchange of letters the arguments on both sides were couched in terms of Einstein's welfare and the advancement of American science. What really mattered was that Flexner was operating from a position of strength. He had the money and was already holding an agreement with Einstein. A few months later, Einstein made his last winter visit to Caltech.[75]

In July, Flexner departed for his annual summer retreat in the Ontario woods. From the isolation of his cabin on Lake Ahmic, he carried out the correspondence with Caltech, Einstein, and Weyl. Flexner's dealings with Weyl were hampered by the primitive state of transoceanic communication, but Weyl's vacillation would have challenged twenty-first century technology. It was early in July that Weyl wrote to express his interest, but not commitment, in the Institute position. One week later Weyl appeared for his appointment in Berlin at the Education Ministry. Weyl proposed the three-year plan that he had discussed earlier with Flexner.[76]

The Minister was disturbed to learn that Weyl contemplated leaving Germany. A three-year trial at the Institute was rejected out of hand. The Minister did react favorably to the concept of a two-year leave for Weyl to visit Princeton. The understanding was that Weyl would remain a Göttingen professor while holding a temporary position at the Institute. No approval was possible until Weyl held a written IAS offer and the proposal was reviewed by the Göttingen scientific faculty. Moreover, the window for consideration was just a few days. Without an offer in that interval, the matter would be put over until November.[77]

Following the meeting at the Ministry, Weyl dined with the Einsteins. The other guests were Institute trustee Frank Aydelotte and Veblen, who both happened to be in Berlin. Weyl explained his predicament, emphasizing the urgency of receiving an offer. Aydelotte cabled the information to Flexner, who had not yet received Weyl's earlier letter.[78]

Flexner immediately cabled an offer to Weyl. The salary and pension figures were identical to those for Veblen and Einstein. An additional education allowance for Weyl's children was to be determined later. Weyl was pleased to receive confirmation in the form he required, but concerned that there was no mention of the two-year arrangement. He cabled Flexner seeking clarification.[79]

Flexner's reaction was uncharacteristic of his conduct in negotiations. He did not immediately reply to Weyl's request. Several factors likely contributed to this decision. Bamberger was becoming impatient with Weyl, perhaps reminded of Birkhoff's questionable tactics. Flexner himself appeared to have mixed feelings over the trial period. With Einstein and Veblen already on board, he could afford to hold off, at least until he received Weyl's follow-up letter. It all became moot when the two-year plan was rejected by the Göttingen faculty. Weyl was being forced to make a choice, a dilemma with which he was ill equipped to deal.[80]

Weyl's next move was to seek enhancement of the Institute offer. Perhaps he could receive some assistance in obtaining a summer home. On the development side, he insisted on a commitment for an additional mathematics position. Weyl reiterated his belief that the current slate of three would benefit from an injection of youth and diversity. He mentioned Artin and John von Neumann as appealing candidates.[81]

Flexner received Weyl's letter in mid-August. He remained confident that Weyl would come around, but was uncomfortable with making a unilateral decision on the new points. Flexner decided, for the first time, to seek the wisdom of the Executive Committee. Following protocol, he requested that the president of the Board, Bamberger, call a meeting. Labor relations was one area where Bamberger had substantial experience. Although Weyl's negotiation ploys were standard procedure for German professors, they did not impress the department store entrepreneur. Bamberger flatly refused to call a meeting and insisted that Weyl decide on the basis of the generous terms that were already on the table.[82]

Flexner was caught in the middle. He wanted Weyl for the faculty, but Bamberger's hardball tactics were no way to hire a sensitive scholar. Flexner searched for a path to make some gesture to Weyl, without a capitulation likely to antagonize Bamberger. Even asking for further consideration forced Flexner into the uncomfortable position of his first direct disagreement with Bamberger.

To make his case, Flexner composed a detailed memo placing Weyl's appointment in the context of establishing the School of Mathematics. He sent this to Bamberger together with a deferential cover letter and the draft of a response to Weyl. Flexner politely asked that the founders discuss the materials with Maass and Leidesdorf. Only if they were fully satisfied was the letter to be sent to Weyl. The proposed modification of the offer was minor, particularly from the financial standpoint. The main enhancements were specific stipends for the Weyl children's education and their support in the event of their parents' death. The commitment to an additional position was particularly weak: "In due time after conference with those constituting the mathematical group the Institute will make every effort to add to the group a younger but promising mathematician, whose main interests lie in a different part of the general field of mathematics."[83] As for the summer home, if Weyl were unable to save a sufficient portion of his salary, then knowledgeable Board members would be happy to advise him in securing a loan.[84]

Bamberger reacted favorably to Flexner's plea for Weyl. Weyl received the revised offer at the end of August. Shortly thereafter the recurring decision paralysis set in. Weyl was unable to choose between Göttingen and the Institute. Late in September, Weyl wrote Flexner asking for more time. Although Flexner would have preferred a decision by the October trustees meeting, he graciously permitted Weyl to proceed on his own schedule. This left Einstein and Veblen to be approved by the trustees as the first two Institute professors. Since their verbal and written offers and acceptances had occurred within days of each other, it is fitting to regard Einstein and Veblen as having equal seniority as the original members of the faculty.[85]

In addition to the professorships for Veblen and Einstein, the trustees approved two other appointments in the School of Mathematics. One of Veblen's Princeton graduate students, John Vanderslice, received a one-year term as Veblen's assistant. Walther Mayer was confirmed as an associate in mathematics. The position of associate was never defined. In fact there did not exist any precedent.[86]

There was a significant distinction in faculty ranks between German and American universities. In Germany there was the bottom level *privatdozent* while the United States had the ranks of assistant professor and associate professor. When Einstein originally negotiated a position for Mayer, it was presumably a professorship that he had in mind. Flexner's written offer designated Mayer as Einstein's assistant. Einstein's counterproposal rejected the assistant classification by seeking independent employment status for Mayer. Flexner's acceptance of these terms and his subsequent written communications about the Mayer appointment repeatedly contained the adjective "independent," but never the term "professor."

When Einstein later sought assurance about Mayer's status, Flexner responded:

> You have no fear respecting Dr. Mayer. He will receive an independent appointment as associate in mathematics on the basis which you yourself suggested and provision for pension will be made for him. I can understand the fact that his scientific standing would suffer if he were regarded merely as your assistant. I hope that the independent appointment may create the impression that he scientifically stands upon his own feet, despite the fact that he is your intimate co-worker.[87]

Flexner's introduction of the term "associate" added little clarity. It had no meaning in either the German or American systems. As the Institute differed from a university, it was reasonable to establish appropriate terminology and job descriptions. In this case, however, Flexner was not creating a new academic status to be filled by Institute scholars. Rather, he was maneuvering to placate Einstein without conferring professorial status on Mayer. Mayer became the first and last associate of the Institute for Advanced Study.

Walther Mayer was born in Graz, Austria, in 1887. He was a mathematician who received his PhD from the University of Vienna, becoming a *privatdozent* in 1926. Mayer is best known for his work with Leopold Vietoris. They developed a basic technique of algebraic topology, now known as the Mayer-Vietoris sequence. It was Mayer's knowledge of differential geometry that, late in 1929, gained him a position as Einstein's assistant. Together they collaborated on the search for a unified field theory. Einstein was so pleased with their progress that he arranged for Mayer to accompany him on the trips to Caltech. Einstein and Mayer published several joint papers.

John von Neumann. (Courtesy of Marina von Neumann Whitman.)

CHAPTER SIX

FLEXNER AND VEBLEN BUILD
THE SCHOOL OF MATHEMATICS

The summer acceptances by Einstein and Veblen gave the Institute its first faculty. With or without Weyl, Flexner was content with the hiring for the School of Mathematics. Two or three professors were the most that he envisioned for any school. It was time to move on to economics and then perhaps the humanities. Altogether Flexner hoped to create as many as six schools over a period of several years.

Although Veblen was well aware of the limited scale in Flexner's plans, he still expected to increase the size of the School of Mathematics. Veblen discussed his ideas with Weyl and included them in his June 5, 1932, acceptance letter to Flexner:

> The point that he [Weyl] emphasizes most is the desirability of having younger men in the group. The names that he and I principally discussed, after Lefschetz, Alexander, and Morse, were Dirac, Artin, and Alexandroff, all under 35. We both admire Miss Noether intensely—she is 50 but still improving—, and think Wiener a serious candidate. He also suggests that there be no distinction between Professors and Associate Professors. This would mean that there would be a group of, say, 7 professors some however at lower stipends. The proposed positions will be more attractive than most American professorships and it would doubtless be confusing to class them as associate professors. Perhaps merely Associates.[1]

Veblen outlined a budget of about $150,000 for the School of Mathematics and named several other candidates for more junior positions.

Flexner's reaction was mixed. He enjoyed Veblen's enthusiasm and its contagious effect on Weyl. However, it was simply unsound to put over half of the Institute's annual income into the first school, especially under the

uncertain economic situation. As for the nominations, Artin had been eliminated. Lefschetz and Alexander were off-limits at Princeton. Flexner sought
to maintain a positive dialogue without making further commitments. He
did indirectly act on one of Veblen's suggestions. Flexner appropriated the
term "associate" for Mayer's ad hoc position.[2]

The seeds of conflict were being sown. Veblen was focused on creating the
best possible mathematics institute while Flexner was determined to maximize
his returns over a diverse range of subjects. In the face of limited resources, it
was inevitable that these worthy objectives would eventually clash. For several
years there was enough money and good feeling to circumvent these difficulties. The first signs of friction emerged when Veblen pushed for a modestly
priced one-year visit by an especially promising young mathematician, Kurt
Gödel.[3]

Three years earlier, at the age of 23, Gödel had completed his PhD at
the University of Vienna. During the intervening period he had obtained
results that revealed astonishing limitations in the foundations of mathematics. Veblen proposed to have Gödel showcase his work at the Institute while
becoming acquainted with America. If Gödel fulfilled his potential, then the
Institute was well positioned to establish a longer-term relationship.[4]

Flexner was negative. "Several considerations would lead me to halt in
connection with Dr. Gödel."[5] Suddenly Flexner was concerned over widen-
ing the scope of the School of Mathematics to include mathematical logic.
There was also the question of Gödel's welfare. "Suppose he comes to America for a year and is not reappointed, where would he be?" These uncharacteristic objections indicate that Flexner was frustrated by Veblen's persistence at
expansion. It was fortunate that their summer separation limited interaction.

Veblen returned from Europe in October. Joining the Institute payroll
brought little change to his routine. He retained his Fine Hall office and
conducted his usual seminar. Flexner continued to work out of his rented
office in New York. Every other week or so, one traveled to meet the other
and discuss the plans for the Institute. A touchy matter was Veblen's campaign
to bring in Gödel and others. Despite Flexner's monumental capacity for
denial, he recognized that there were problems in his relationship with Veblen.
Insecurity about his own lack of scholarly accomplishment resurfaced. Abe
looked to his brother Simon to straighten out Veblen.

Simon understood both Veblen and Abe. He appreciated the nature of
their conflict and knew that Abe was unrealistic in expecting a magical resolution of the problem. Although Simon was reluctant to be placed in the

middle, he was unable to refuse his younger brother. The three met on October 20, ostensibly to discuss the workings of an institute. Simon explained the operation of the Rockefeller Institute. Veblen was particularly interested in the roles of the Director and the scientific staff in setting research directions and selecting personnel. He noted that Simon's scientific qualifications placed his directorship in a different category from that of a typical administrator.[6]

On the following day, Abe wrote to Simon declaring victory.

> What you said and what you showed Veblen yesterday greatly cleared his mind. I think he sees now the difference between a university department and the sort of thing that has been in the back of my mind, based so largely on what you have done. Important also is the fact that he was convinced of the necessity of slowly feeling our way ahead rather than trying to start out with a considerable group in any one subject.[7]

Simon was less sanguine. He observed that Abe's name dropping, so effective on philanthropists, grated on Veblen. Simon advised, "I would not quote the Rockefeller Institute to him any more."[8] Abe would fail to heed Simon's advice. To Veblen, and everyone else, Abe invariably supported his positions by invoking the experience of his brother at the Rockefeller Institute.

Despite their differences, Veblen and Flexner were an effective team. Veblen had a keen sense for recognizing mathematical talent. He performed the screening and Flexner made the final decision. Out of this sometimes acrimonious process came a number of extraordinary appointments. The first breakthrough was on Gödel, where the circumstances were tangled with a landmark decision on Institute policy. The issue involved the admission requirements for Institute students.

Flexner's original plan was to follow the Gilman model for a graduate university. While there would be considerable flexibility for professors in choosing their disciples, a typical student was to arrive with a bachelor's degree and then work toward a PhD. Around October 1932 Flexner acted unilaterally to stiffen the entrance requirements. A PhD became a prerequisite for admission. Plans to award degrees were abandoned. The success of Johns Hopkins in lifting American graduate education had made it possible to take this next logical step. Flexner was moving beyond Gilman. The Institute need not waste time on the formalities of degree requirements, administration, and awards. There would be a total emphasis on research, by individuals who had already established their commitment and capacity. Promising young PhDs were to receive the final direction and seasoning at the Institute before moving on to positions at American universities.[9]

Under this revised structure Flexner saw a place for Gödel. In November Flexner wrote Veblen:

> We now have in you and Einstein two persons certain. I would suggest that, dropping for the time being questions of site, etc., you begin to get together two or three or four men to work with and under you, beginning next autumn. You have, I suppose, already written Gödel, and you mentioned Whitney the other day . . . . Should Weyl accept, we shall treat him similarly.[10]

Veblen had no intention of working with or over Gödel, whom he hoped would collaborate with another Princeton mathematician, Alonzo Church. Veblen's priority was to bring the best research mathematicians to the Institute. The new postdoctoral aspect was well suited to achieving the objectives of both Veblen and Flexner. At this stage one difference in their thinking was that Flexner viewed these young people as students while Veblen regarded them as independent researchers. In any event the postdoctoral policy became a distinguishing feature of the Institute, forever setting it apart from Johns Hopkins. Gödel received a one-year appointment with a $2,500 stipend to an unspecified position. It was the beginning of his long association with the Institute. Today Gödel's work is widely regarded as one of the seminal intellectual accomplishments of the twentieth century.[11]

Flexner remained oblivious that, in Gödel, he was obtaining precisely the type of profound, original thinker for which the Institute was created. Still, it was clear to him that Veblen was the talent scout and sounding board on whom he had come to rely in the successful launch of mathematics. To apply the template to economics required a leader such as Veblen. Meanwhile the appointment of Einstein had set a standard that was impossible to replicate.

With the October 1932 Board meeting came the public announcement of the Einstein and Veblen appointments. It was a front page story in the *New York Times*. Congratulations flowed to Flexner from everywhere. Among the well wishers were his social science consultants Charles Beard and Felix Frankfurter. Beard included a *mea culpa* for his previous advice to "chuck mathematics." Earlier that year, Flexner was disposed to appoint Beard to head a school in the social sciences. Now he shifted to Frankfurter as the putative choice for economics.[12]

It is easy to understand the appeal of the astute 50-year-old Harvard law professor. Frankfurter had been an advisor to President Woodrow Wilson and was currently performing that role for President-elect Franklin Roosevelt.

Frankfurter's students were assuming positions as Supreme Court clerks and government officials. The combination of academic and non-academic connections was what Flexner was seeking for economics. Moreover, the influential Frankfurter was already Flexner's most trusted advisor on economics appointments. The big question was whether Frankfurter would leave Harvard to join the Institute. Unlike Beard, who supported himself by book royalties, Frankfurter occupied one of the most prestigious positions in America. Only the Supreme Court could offer more compelling circumstances for his intellect.

Frankfurter had shown genuine enthusiasm for the Institute during its conceptual stages. Perhaps he would be interested in joining Einstein on the faculty. Frankfurter's acknowledgement of the Einstein appointment and their continuing discussion over identifying personnel in economics gave Flexner an opening. Early in November he concluded a letter with an offhand overture: "You, yourself, may be the fellow."[13] As Flexner waited for Frankfurter's reaction and continued the vigil for Weyl's decision, he turned to the delicate issue of relations with Princeton University.

Several Princeton area sites were under consideration for an Institute campus, but to obtain a bargain it was prudent to wait for the right opportunity. No purchase was imminent and it was clear that the School of Mathematics would remain in Fine Hall for the foreseeable future. While cohabitation with the University mathematics group provided attractive opportunities for collaboration, there was also a risk of friction. The discrepancies in Institute and university salaries, pensions, and teaching duties were becoming known. It would be natural for the more favorable circumstances of the Institute faculty to arouse envy among their Princeton colleagues. Would the Institute take additional faculty from the university? If so, what would be the impact on the Princeton mathematics department? Flexner's original planning had excluded university faculty from consideration. It was more or less out of desperation that Flexner had turned to Veblen. Flexner viewed the action as an exception, but Veblen was pushing for Alexander and Lefschetz to follow him. The prospect of additional transfers was an especially sensitive matter.

Flexner sought to clear the air with Eisenhart. Mindful that the precedents of the current time were likely to bind their successors, Flexner knew that it was essential to have a thorough vetting of policy. Paramount in Flexner's thinking was that the Institute and university be viewed as independent operations. The difficulty was that if the Institute acted now, purely in its own interest, then it would pursue Veblen's desire to raid the Princeton faculty. Flexner referred to this as "moving men from one place on the checkerboard to another." To strengthen the Institute at the cost of weakening its host was not only ungrateful but failed to advance the strategy of making over the city of Princeton into the mathematical center of the universe.[14]

Under Flexner's analysis, this conflict existed largely in the short run, during the Institute's formative period. The solution that he proposed was a strict, formal policy of independence, accompanied by a secret, temporary moratorium on transfers. Flexner emphasized to Eisenhart that as long as they each headed their respective institutions, further accommodations presented no problems. Eisenhart was at this time functioning as the chief academic officer of Princeton University. Hibben had retired from the presidency a few months earlier, and the chair of the trustees was serving as acting president from his base in Newark. Eisenhart was both dean of the faculty and chair of the mathematics department. A search was underway for a new president and Eisenhart was receiving strong consideration.[15]

Eisenhart's reaction to the proposal was unexpected. Rather than accept protection from Institute poaching, he worried that any de facto moratorium might hinder university recruiting. If a promising scholar aspired to an Institute professorship, then he or she might be advised to avoid employment at the university. On November 26, 1932 Eisenhart wrote Flexner:

> I agree with you that the relationship of the Institute and our Department of Mathematics must be thought of as a matter of policy extending over the years. Accordingly I am of the opinion that any of its members should be considered for appointment to the Institute on his merits alone and not with reference to whether for the time being his possible withdrawal from the Department would give the impression that such withdrawal would weaken the Department. For, if this were not the policy, we should be at a disadvantage in recruiting our personnel from time to time. If our Trustees and alumni were disturbed by such a withdrawal, as you suggest, they should meet it by giving us at least as full opportunity to make replacements intended to maintain our distinction. The only disadvantage to us of such withdrawals would arise, if we were hampered in any way in continuing the policy which has brought us to the position we now occupy. This policy has been to watch the field carefully and try out men of promise at every possible opportunity. If it is to be the policy of the Institute to have young men here on temporary appointment, this would enable us to be in much better position to watch the field.
>
> In my opinion the ideas here set forth are so important for the future of our Department that it is my intention to present them to the Curriculum

Committee of our Board of Trustees at its meeting next month, after I have had an opportunity to discuss them further with you next week.[16]

Eisenhart was seeking affirmation for the policy, as well as assurance that the mathematics department would be funded to replace scholars moving to the Institute. At the December 18 meeting of the Curriculum Committee, Eisenhart's presentation was "approved in principle." The "in principle" stipulation may have been intended to leave room for flexibility in responding to the circumstances of each individual resignation. It is unclear whether any controversy arose in the discussion. One trustee followed up with a letter of dissent. He predicted that Eisenhart's policy would doom the Princeton mathematics department to a supporting role for the Institute.[17]

Flexner appreciated the concern. He really intended to appoint outsiders such as Birkhoff, Einstein, and Weyl. Bringing them to Princeton enhanced science at the university in a fashion that would have been impossible without the existence of the Institute. It was an additional reason that Weyl was important. Weyl had led Flexner to expect his decision in late October. When November arrived, Flexner and Veblen were still checking their mail each day looking for a Weyl response.

It fell to Veblen to send a diplomatic prompt to Weyl. Veblen gave an upbeat progress report on the Institute, pointing out that major moves were waiting on Weyl's acceptance. This note crossed with a long, rambling letter from Weyl. Weyl spoke as if he were coming to Princeton, saying there was at most a 15% likelihood of his remaining in Göttingen. Typically, there was no definitive decision.[18]

Flexner sought to coordinate his response with Veblen's,[19] but there was little more that either could say. Veblen mainly confined himself to the questions raised by Weyl. Flexner emphasized the unique opportunity afforded by the Institute:

> There is only one additional comment that I should make, namely—it occurs to me that, in thinking of the Institute and particularly the School of Mathematics, you may have had in mind something like a university faculty in mathematics. This seems to me a misconception, for it is precisely the advantage of the Institute that, unlike a university faculty, there is nothing it must do unless really first-rate persons can be obtained.[20]

This echoed a sentence in the letter that Veblen had sent earlier in the month:

> In an ordinary American university it is easy to use a person who does not live up to his early promises for teaching and administration, whereas in the Institute the scientific work is the only thing.[21]

It was all innocuous stuff with which Flexner and Veblen were giving Weyl the space to make up his mind. Weyl, however, was undergoing a mid-life crisis. Weyl feared that he was beyond his mathematical peak and would be unable to sustain his flow of groundbreaking results. He knew that Hilbert's great work on integral equations was created between the ages of 42 and 46. Weyl, who himself was turning 47, judged Hilbert's subsequent output as infrequent and inferior. There was also the example of Klein. Klein's greatest contributions to Göttingen were as a teacher and recruiter of Hilbert. His best mathematics was done as a young man prior to his arrival.[22]

At German universities a professorship in mathematics was a reward for past accomplishments. Professors had substantial freedom to make their contributions through some combination of research and teaching. Weyl, who did not care for one-on-one supervision of students, had a desire to impart his profound mathematical perspective through lectures. Weyl was actually hoping for some assurance that he would not be expected to produce great theorems every year. The statements by Flexner and Veblen unwittingly threatened to work against their cause.

Weyl's self-doubts were added to his many other considerations. As for the conditions that were driving Weyl from Germany, the developments were mixed. The Nazis actually lost ground in the November 1932 election. This gave a deceptive impression that Hitler's influence had peaked. Meanwhile the economy was in the depths of its long-term decline. Statistics on industrial production, the stock market, and employment were abysmal. The government was in financial trouble and there was little flexibility for the Education Ministry to respond with counteroffers. When Weyl followed the custom of seeking to leverage improvements in his circumstances at Göttingen, he was flatly refused.[23]

Weyl was feeling the pressure from both sides to reach closure. The negative response from the Ministry temporarily ended the indecision. On December 1, just before the arrival of Flexner's November letter, Weyl cabled and wrote Flexner with his intention to accept the Institute position. When Flexner's letter arrived, Weyl was gripped by fear. What if his research career were over? At Göttingen he could still be productive. In fairness to Veblen and Flexner, it should be pointed out that they held hopeless positions. Weyl's paranoia could have even been triggered by the salutation in their letters.[24]

Flexner had no reason to doubt Weyl's acceptance. All along he had been confident that Weyl would come around to realizing that the Institute was in his best interest. It had been six months since the acceptances of Einstein

and Veblen. During that period Flexner's feelings of accomplishment were
punctuated with anticipation over the prospect of adding Weyl. With Weyl
finally on board, Flexner had succeeded beyond any reasonable expectation.
Einstein, Veblen, and Weyl gave the Institute for Advanced Study a level of
scholarly pre-eminence unprecedented in America. Princeton would displace
the depleted Göttingen as the foremost mathematical center in the world.
Flexner had actually achieved the ambitious objective that he had set for his
first school.

At this point it should have been easy to predict the moves of Veblen and
Flexner. Veblen could be expected to lobby for additional positions. Flexner
was likely to turn his attention to economics or another school. He could
make the reasonable argument that further mathematics personnel decisions
be deferred until the arrival of Weyl and Einstein the following fall. What
mattered most was that the power rested entirely in Flexner's hands.

Surprisingly, Flexner was unable to feel contentment with the impressive
mathematics roster. It had less to do with Veblen's campaign, but was a matter of demographics. Despite the Institute's founding ideals of inclusiveness,
Flexner was uncomfortable with the German majority on his mathematics
faculty. If he could add just one American, there would be a perfect balance.
There should not have been any hurry. After all, Flexner was committed to
including Weyl in any subsequent personnel discussions. Perhaps Weyl's dal-
liance had forfeited his seat at the table. In any event Flexner would never
understand the importance to his faculty of having a voice in the selection
of their colleagues. He decided to bypass Weyl and act quickly. The timing
here is important. Weyl's acceptance arrived between the time that Eisenhart
had proposed a policy of complete independence and the scheduled meeting
where the Princeton Curriculum Committee was to ratify the principle. If
another university mathematician were to be appointed, the time would be
ripe after the meeting.[25]

There was a major logistical advantage to recruiting Princeton faculty. A
transfer from the university to the Institute did not even require a change
of home address. The decision process should be short and avoid many of
the issues that plagued Birkhoff and Weyl. This further suited Flexner's haste
to complete the mathematics faculty. The conflict over checkerboard moves
was reconciled with the urgency of raising the domestic factor in the faculty
nationality ratio.

The leading American candidates from prior discussions were Alexander
and Lefschetz of Princeton and Marston Morse of Harvard. Each was a ge-

ometer who was being pushed by Veblen. Veblen ascertained that Morse was inclined to remain at Harvard. Weyl, who had privately expressed doubts to Flexner about Lefschetz, preferred a non-geometer. Weyl's August offer even included a commitment that the next appointment be in an area that was not already represented. Flexner neglected this stipulation and began positioning himself to secure Alexander.[26]

James W. Alexander II was born in Sea Bright, New Jersey, on September 19, 1888. He came from a household of wealth and culture. His father was the eminent painter John White Alexander. On the maternal side his grandfather James W. Alexander was president of the Equitable insurance company. The mathematician's mother, Elizabeth Alexander Alexander, was a prominent patron of the arts. Family friends included Henry James, James Whistler, Auguste Rodin, and other famous, creative figures of the time.[27]

The future mathematician's childhood was spent in Paris as his father's artistic standing was rising to international renown. The family returned to the United States in 1901. Five years later Alexander matriculated as an undergraduate at Princeton University, completing his degree in 1910. He stayed on at Princeton to pursue graduate work in mathematics. Alexander was an unusual graduate student in that he followed two different research directions. His thesis, under the supervision of T. H. Gronwall, was an important contribution to complex analysis.[28] Meanwhile he was guided by Oswald Veblen into the emerging area of analysis situs (algebraic topology). Alexander and Veblen collaborated to shore up the foundations that had recently been erected by Poincaré. It would not be long before the pupil surpassed the mentor. This chronology has left some with the mistaken impression that Veblen was Alexander's thesis advisor.

Alexander made the transition from graduate student to faculty member in the Princeton mathematics department. He became a full professor in 1928. Some indication of Alexander's importance to topology is given by the list of structures that bear his name: Alexander duality, Alexander horned sphere, Alexander polynomial, and Alexander-Spanier cohomology. Not among these was his first major result. Algebraic topology is an interdisciplinary field with algebraic tools being employed to identify topological differences. Poincaré's approach was to take a topological object, now known as a manifold, and decompose it into more basic pieces, called a cellular decomposition. The algebraic features were the Betti numbers and torsion coefficients obtained from the decomposition. For the program to make sense, it was vital that employing different decompositions of the same manifold would result in the same Betti numbers and torsion coefficients. In 1915 Alexander obtained this fundamental theorem.[29]

When Flexner considered Alexander at the end of 1932, algebraic topology was still a young subject. Flexner could not have foreseen that it would become one of the most fruitful mathematical areas of the mid-twentieth century. Alexander's pioneering results played a major role in permitting the subsequent development. In recent years the Alexander polynomial has enjoyed renewed influence. It arises in the subspecialty of *knot theory*. Mathematical knots are determined by simple closed curves in space. Alexander discovered a way to associate a polynomial to each knot. These polynomials were surprisingly (but not totally) successful at distinguishing topologically different knots. Over the past two decades Alexander's work has served as the starting point for a more comprehensive approach.

In a letter to his daughter, Flexner summarized what he knew about Alexander the mathematician: "He is just past forty and in Veblen's judgement is the best bet in this country at that age."[30] Flexner found other aspects of Alexander appealing, including his politics. While Flexner himself was a moderate, he respected a range of left-wing perspectives such as those of Alexander, Veblen, and Beard. Another significant factor was Alexander's cosmopolitan orientation. In addition to his early education in France, Alexander was married to a Russian woman, Natalie Levitzkaya. The couple summered in the village of Les Houches near Mont Blanc.

The Alpine location suited Alexander's avocation. He was an accomplished mountaineer, taking up the activity on a 1921 trip to Colorado. The following year he made a first ascent along an icy route to Long's Peak in the Rocky Mountain National Park. This climb and one of its structures are now known as Alexander's Chimney in his honor. It was the access to the Alps that attracted Alexander to Les Houches, mixing mathematical research with the challenging mountain routes.

Alexander's worldly lifestyle was made possible by his independent wealth. This provided him with options that were not available to other scholars. When Alexander decided that teaching was taking up too much of his time, he negotiated a reduced course load by forgoing half of his salary. The financial aspect was another consideration for Flexner who, for the first time, was feeling the constraints of a budget. Alexander would be available for considerably less than the $15,000 figure being paid to the others. Flexner had no qualms over introducing a disparity among the staff. He firmly believed that salary was appropriately linked to need.[31]

Flexner moved quickly to consolidate his position on Weyl and Alexander. When Weyl's telegram arrived on December 2, it included a request for a binding contract. The next Board meeting was over a month away. However, on the following week there was an already scheduled luncheon for Flexner to update the founders and their advisors on Institute developments. It was a simple matter to convert the luncheon into a meeting of the Executive Committee. At this gathering the Weyl appointment was ratified and Flexner was authorized to negotiate with Alexander. This empowered Flexner to write Weyl with an unqualified offer of the terms that he had just accepted, in principle.[32]

The Institute trustees were serving in a largely ceremonial capacity. Flexner ran the show. He consulted Simon, Veblen, Frankfurter, and others as needed. When his mind was made up, he sought approval from Bamberger. Maass was often in on the briefing and sometimes Fuld, Leidesdorf, and Aydelotte. The other trustees rubber stamped the decisions. It was not the type of governance that Flexner had advocated in his book, but Bamberger had insisted on seating his cronies. Flexner had little difficulty adapting to his role of autocrat.

It was hard to criticize the results. A first school with Einstein, Veblen, Weyl, and possibly Alexander was an amazing realization of the Institute mission. Bamberger and Fuld were so satisfied with the progress that they announced their intention to retire from their formal positions on the Board, ostensibly leaving the operation of the Institute in the hands of the Director and trustees. It was agreed that the founders' titles of President and Vice President would be exchanged for Honorary Trustees. The main effect on them was that future attendance at Board meeting would be welcomed, but not expected. [33]

The departure of the founders called for a reorganization of the Board. Only Alanson Houghton remained in a leadership office. Houghton, who held the title of Chairman, functioned as a figurehead. Flexner needed to fill the vacuum of command with someone who could serve as liaison between Bamberger and himself. The clear choice was Herbert Maass. Among those in Bamberger's inner circle, Maass was more engaged than either Hardin or Leidesdorf in following the intellectual development of the Institute. Moreover, Maass and Flexner had established a cordial working relationship over the past three years. Maass became Vice Chairman, and the titles of President and Vice President were (temporarily) retired with their original holders.[34]

The by-laws vested the Board with considerable authority. The extent to which it was exercised was largely in the hands of the executive officers.

Neither Bamberger nor Houghton were disposed to have the trustees play a substantive role. Nevertheless the power was there to take actions such as rejecting faculty appointments or deposing Flexner. It would be many years before Maass would flex his muscles, but the new status opened his path to power.

Even as an Honorary Trustee, Bamberger retained the most important leverage. This was the prospect that he would transfer more of his wealth to the Institute. The original $5 million contribution provided ample endowment for the first few years, but additional funds were essential to complete the project. From the start Flexner was confident that he could cultivate Bamberger to deliver at the appropriate time. Under this strategy Flexner would continue to clear his actions with Bamberger.

Restructuring of the Board was just one of several issues that occupied Flexner in mid-December. His hopes for the social sciences received a blow with the arrival of a letter from Frankfurter. There was no response to the overture about joining the faculty. The door was not closed, but it was again time to revise the plan for the second school, or abandon it altogether.[35]

The immediate opportunity was in mathematics. Flexner was anxious to complete the School's faculty with a second American. The trustees had given him clearance on Alexander. The University Curriculum Committee of trustees was scheduled to consider Eisenhart's proposal on December 17. While this would clear the way to approach Alexander, Flexner remained conflicted over whether this move would be disrespectful to Eisenhart and Princeton. If only Eisenhart could have some voice in the decision.

Alexander was not the only prominent geometer at Princeton. Solomon Lefschetz, a naturalized American citizen, was also in Veblen's top tier. In fact, it was Alexander and Lefschetz who were the two candidates to succeed Veblen as Fine Professor. Although the Institute position was regarded as more desirable, Flexner decided to give Eisenhart first choice on whom to retain. It was a plan that was courteous to the university and guaranteed the Institute its second American geometer. The reasoning process also exposed certain values of Flexner. National origin was unimportant but citizenship was crucial.[36]

Solomon Lefschetz was born on September 3, 1884, in Moscow, Russia. His Jewish family settled in Paris, France, where he was raised and educated. Lefschetz was trained as an engineer. In 1905 he came to the United States to pursue a career in industry.[37]

Tragedy struck in 1907 when Lefschetz lost both of his hands in a laboratory explosion. Confronted with a dependency on the prostheses of the

Solomon Lefschetz, 1935. (Courtesy of the American Mathematical Society.)

day, Lefschetz made the pragmatic decision to change careers. In 1910 he entered Clark University and quickly obtained a mathematics PhD under the direction of William Story. One can only imagine the employment discrimination then facing a Russian-born Jewish mathematician with a new PhD and a severe disability. Lefschetz found an instructorship at Nebraska and then moved to Kansas at the same rank in 1913.

Despite the additional obstacles of a high teaching load and mathematical isolation, Lefschetz' research flourished. He succeeded in applying the techniques of Poincaré to resolve fundamental questions in algebraic geometry. This work received international acclaim when the prestigious Paris Academy of Sciences awarded Lefschetz the 1919 Bordin Prize. The only immediate career reward for Lefschetz was a promotion from assistant to associate professor. Lefschetz remained at Kansas for the next five years.

In 1924 Lefschetz came to Princeton in a visiting faculty position which was made permanent the following year. During this time he formulated the famous Lefschetz Fixed Point Theorem. This gave algebraic topological conditions on a function and manifold that can guarantee the existence of a fixed point. The result and its refinements have had an enormous impact on twentieth-century mathematics.

Portrayals of Lefschetz depict a mathematician of awesome power. It was said that his mathematical senses were so keen that he seemed to be able to smell any theorem in his vicinity. Lefschetz always focussed on the big picture and had little patience for details. His lectures were often incomprehensible. On the personal side Lefschetz was outspoken and abrasive.[38]

The choice between Alexander and Lefschetz presented striking disparities in background, approach, and temperament. Alexander was from privileged circumstances while Lefschetz had overcome formidable barriers. Alexander was an elegant mathematician with an urbane manner. The coarse Lefschetz aggressively attacked his subject. The principle similarities were their Paris upbringing and magnificent geometric theorems.

It was up to Eisenhart to decide who would remain at Princeton. Regardless of his selection, both Alexander and Lefschetz were assured a salary increase to $10,000. Ironically, the winner of the competition was actually the loser in that he would not receive the preferred position. In many respects Alexander was the obvious choice. Princeton was a traditional university where both inbreeding and anti-Semitism were prevalent. Not only had Alexander been a Princetonian for a quarter century, but his maternal ancestors figured prominently in community history. Campus buildings and a town road carried the Alexander name. This legacy contrasted sharply with that of the Jewish outsider, Lefschetz.

Eisenhart, however, had additional considerations. The mathematics department had already lost Fine and Veblen. Eisenhart himself was now occupied with running the university. The department was in need of a forceful mathematician to lead it over the next decade. On these grounds the driven Lefschetz was a better bet than the half-salaried Alexander. Eisenhart pondered the merits of both candidates and decided to retain Lefschetz. Everything then fell quickly into place. Eisenhart's policy proposal was approved, and Alexander received an offer to join the Institute. Acceptance was a formality.[39]

Flexner's machinations reveal his insensitivity to the feelings of faculty. Flexner believed that dedicated scholars were only concerned with finding the best conditions for their research and family. Since the Princeton and Institute positions carried identical remuneration and working environment, there was no rational reason for Lefschetz or Alexander to care whether they were Fine Professor or a member of the Institute faculty. Lefschetz did care. The injury that he felt from the process undoubtedly contributed to the bitter attitude that he would later exhibit toward the Institute.[40]

Completing the mathematics faculty provided a satisfying end to 1932 for Flexner. Writing to his daughter, he commented: "I rounded out my mathematical group by asking Alexander, one of the Princeton mathematicians, to take the remaining professorship."[41] Regardless of the expectations of Veblen and Weyl, no additional hires were programmed for mathematics. The eminent standing and national balance of the faculty were all that Flexner desired. Moreover, the age distribution guaranteed stability: "I have now an age group of four men ranging from 40 to 52, so that the whole chebang cannot go to pieces at once." The Institute mathematics program would not suffer the fate of Hopkins when it lost Sylvester.

Nevertheless, it was a fragile group. Arrival of the Germans could never be taken for granted. Einstein was such an important world figure that there was always the possibility of some miraculous new opportunity becoming available to him. Meanwhile Weyl was under siege by the anxiety that always accompanied his career decisions. Had he acted in the best interests of his family? Could he continue to do first-class mathematics? What about his responsibility to upholding the Göttingen tradition? Weyl remained haunted by his mistake in leaving Zurich. Then he had waited too long to undo the decision. A reversal on Princeton would have to be undertaken more quickly. Weyl was fully cognizant that some leeway still existed. He had not yet resigned from Göttingen nor had he acknowledged his acceptance of the December terms authorized by the Institute Executive Committee.

The vacillation in Weyl's mind was captured in a bizarre sequence of three cables to Flexner during the first week of 1933. On Monday night Weyl sent a message with his acceptance of the Institute terms subject to an added stipulation. He asked that the pension arrangement be understood to cover early retirement in the event of disability. By the following morning he had changed his mind. Weyl cabled that he was unable to leave his homeland. On Wednesday night Weyl again reversed course, this time accepting irrevocably ("unwiderruflich") the Institute offer.[42]

The Weyl cables must have given Flexner flashbacks to his dealings with Birkhoff. An eminent mathematician was behaving erratically and in a questionable manner. Once again Flexner maintained his magnanimity. He replied to both the second and third cables with understanding and complete support.[43]

Amidst Flexner's calm assurances to Weyl, significant contingency activity was taking place behind the scenes. Flexner received the second Weyl telegram on Tuesday and relayed the information to Veblen. When the third telegram arrived at the Institute's New York office on Thursday, Flexner was out of town. It was on Friday in Princeton that Flexner learned of Weyl's *irrevocable* acceptance. In the intervening three days Flexner and Veblen faced the new reality of a School of Mathematics faculty without Weyl. A trustees meeting was scheduled for the following Monday and the Weyl appointment was on the docket. Flexner was confronted with some serious damage control. It would be difficult to explain Weyl's behavior in a way that did not reflect badly on the Director who recruited him.[44]

Veblen saw an opportunity to push for a new appointment. His choice was exquisite. Veblen suggested filling the Weyl void with John von Neumann, a Jewish Hungarian mathematician holding a temporary position at Princeton. It was well known that the brilliant von Neumann desired a more permanent arrangement in the United States. With him already located in Princeton, there was a reasonable expectation of ascertaining his intentions prior to the Monday trustees' meeting. This new candidate could provide Flexner with an upbeat segue to minimize the embarrassing report on Weyl.

Other features of von Neumann were well suited to quick action by Flexner. Flexner placed great importance on his personal assessment of a potential faculty member. As they were already acquainted, Flexner had observed the charm and intellect that were hallmarks of the impression that von Neumann made on everyone. Moreover, von Neumann, who had just turned 29, would demonstrably improve the faculty age distribution. Even the nationality issue worked in von Neumann's favor. The position was already programmed for a European. Unlike Weyl, who was conflicted over living in the United States, von Neumann had commenced the process toward naturalization.[45]

Flexner held one serious reservation over making the appointment. Von Neumann was also under consideration by Eisenhart to fill one of the professorships that had been vacated at the university. Flexner and Eisenhart were in agreement that it would be desirable to attach von Neumann to one of the Princeton institutions. While both had vacancies, it was clear that Flexner was in a position to act more quickly and offer a higher salary.

The situation changed on Friday. Word of Weyl's acceptance reached Flexner prior to his meeting with von Neumann. Suddenly the question was whether to create a fifth mathematics professorship or to yield to the university. Flexner deliberated the issue as he spoke with von Neumann. Unable to reach a decision, Flexner hedged on offering an Institute position.[46]

Veblen then went to work on Flexner. An obvious consideration was the possibility that a fourth cable from Weyl could materialize at any instant. When Flexner left Princeton on Friday, he seemed inclined to make the von Neumann appointment. Certainly it was permissible within the university-Institute recruitment understanding. Returning to New York, and away from Veblen, a sense of guilt set in. The university needed von Neumann more than the Institute. Flexner was indebted to Eisenhart, whose principled and selfless cooperation were so crucial to the Institute success. On Sunday evening Flexner and Eisenhart discussed the matter over the phone. In spite of their formal policy of independence, the two administrators agreed that the Institute would step aside on von Neumann and leave him for the university.[47]

When Flexner reported to the trustees on Monday, it was as if the previous week had not occurred. There was no mention of von Neumann nor of the collusion with the university. Flexner merely stated, "I cannot sufficiently express my gratitude to the authorities of Princeton University, who have whole-heartedly cooperated with me in the endeavor to make our group as strong as possible."[48] Flexner's portrayal of Weyl omitted any reference to his bizarre antics.

No cables arrived from Germany on Monday or Tuesday. It must have been a relief for Flexner that Weyl was observing a period of transatlantic silence. It was not because Weyl had overcome his demons. His personal state actually deteriorated after sending the third telegram. Weyl's mother-in-law died suddenly and then the entire household (family and servants) contracted influenza. It was more than Weyl could handle. He suffered a mental breakdown. The prospect of moving his family to America became overwhelming. Weyl cabled Flexner, begging to be released from the commitment he had made one week earlier.[49]

Flexner was shocked. He concluded that Weyl was too high maintenance to allow the offer to remain on the table. The exasperation was apparent in the following notification of closure:

> I confess that I was amazed by your cablegram of yesterday following the cablegram in which you committed yourself "irrevocably". On the other hand, I have not the slightest desire to hold you against your will and wish. Our negotiations are therefore terminated and cancelled, and I have so cabled you.

> Inasmuch as the Board ratified your appointment, we shall have to go through the formality of accepting your resignation, which is a mere matter of record, but permits no subsequent change of mind on your part. This can be done when the Executive Committee meets in the near future.[50]

Weyl was in no condition to react. Paralyzed by depression, he was on his way to a sanitorium. The timing of his January 11 declination, however, deserves a place among the fateful decisions of the period. A German man with a Jewish wife burned his bridge out of the country. On January 30 came the surprise announcement of Adolf Hitler's appointment as chancellor. Weyl's future in Germany was doomed.

Flexner found himself in the contingency that he considered two cables and one week earlier. In the interim the entire Board had ratified the Weyl appointment and Eisenhart was moving to offer von Neumann an $8,000 mathematics professorship. There was no time to waste. Flexner quickly decided that he wanted von Neumann as a replacement for Weyl. The first step was to smooth things over with Eisenhart.[51]

In the morning after receiving Weyl's telegram, Flexner met with von Neumann.[52] Agreement was reached on an Institute professorship with the same $10,000 terms as Alexander's. It only remained to obtain the rubber stamp of the trustees. Reaching Aydelotte was especially urgent in that he was completing an article on the Institute. Any mention of the Weyl hiring had to be suppressed. Flexner gave careful consideration over how to explain the recent personnel developments. He followed a strict policy of supporting his faculty to the trustees. However, there was no longer any need to protect Weyl. In his letter to Aydelotte, Flexner rehearsed his trustee version of events. Weyl's vacillation revealed that he was "neurotic" and a "fool."[53] Flexner went on to claim that he was not entirely taken in. Von Neumann was someone "I all along had up my sleeve" as a replacement. Flexner neglected to mention that the von Neumann ace had been firmly planted by Veblen.[54]

The next trustee meeting was scheduled in April. Expunging Weyl from the record called for quicker action. Flexner again converted a luncheon into an Executive Committee meeting. At the end of the month Flexner reported to the founders, Leidesdorf, and Maass. The official minutes describe how Weyl's health precluded a move to a new environment. "After discussion" the appointment was annulled. It is likely that this unrecorded discussion involved the details of Weyl's behavior. The privileged information would be kept confidential in the interests of Weyl and the Institute.[55]

The Executive Committee then approved the appointment of John von Neumann. Like Einstein, the 29-year-old von Neumann would become one of the most significant scientific figures of the twentieth century. It was another landmark hire for the Institute and a curious consequence of Weyl's feeble decisions.

Von Neumann was born December 28, 1903, in Budapest with the name Neumann Janos. Transposition and John(ny) were Americanizations. *Von* was a title of nobility that the family acquired ten years after his birth and, indeed, John's circumstances were privileged. His father was a well-educated lawyer who represented a leading Hungarian bank. His maternal grandfather acquired considerable wealth in the agricultural equipment business. During his upbringing John lived in a spacious apartment surrounded by extended family and servants. [56]

John and his siblings were raised as non-observant Jews. At the turn of the century Budapest offered an unusual array of professional opportunities to the Jewish community. Moreover, Budapest was among the cultural and educational centers of the world, making it an ideal environment for the remarkably precocious von Neumann. Even as a child John showed exceptional skill in languages, mathematics, history, and other subjects. Tutors were engaged to nurture these talents.

In Budapest, institutional education began at age ten. John was among the many Jewish students enrolled at the Lutheran Gymnasium. It was an excellent school. One year ahead of John was Eugene Wigner, who became a lifelong friend and would win a Nobel Prize in physics. Even amidst this competitive environment von Neumann stood out as a prodigy. In addition to following the Gymnasium curriculum, von Neumann was sent to Budapest University for individualized direction in mathematics. There, as at every juncture of his life, von Neumann displayed his extraordinary intellectual power and drive. Despite his adolescence von Neumann rapidly assimilated into the rigorous Budapest mathematical culture.

In a von Neumann biography Wigner recalls asking his younger schoolmate how to prove an established theorem in number theory.

> To do so, said Johnny, it would be useful to invoke certain other theorems in number theory. Did Wigner, to begin with, know such-and-such a theorem? Wigner did. Did he also know such-and-such a theorem? Wigner did not. The catechism continued. Wigner, it turned out, knew quite a few of those useful theorems but was ignorant of others. Johnny paused for a few moments of deep thought. Using only the subsidiary theorems with which Wigner was acquainted, he then provided a proof of the theorem with which the discussion had begun.[57]

When, at 17, von Neumann graduated from the Lutheran Gymnasium, he was already a serious mathematician. Von Neumann was submitting his first paper for publication and would win a competitive mathematics examination for a prestigious national prize. A family decision was then made for

John to prepare for a more lucrative career in chemical engineering. This program would take him to Berlin and Zurich. Meanwhile he was permitted to continue his passion for mathematics with a sort of correspondence study at Budapest.

Von Neumann became interested in one of the central mathematics problems of the day. This was to identify a set of axioms that provide an effective foundation for mathematics. It was a dear problem to David Hilbert and among those he posed in his famous list to challenge mathematicians of the twentieth century. Von Neumann's investigation led him to discover the definition of *ordinal* that remains in use today. His larger progress in the area came to the attention of Hilbert. This resulted in an audience for von Neumann with Hilbert in Göttingen.

In his early twenties von Neumann's mathematical prowess was already well known among Europe's leading mathematicians. While studying chemical engineering at ETH in Zurich, he interacted with Hermann Weyl in the mathematics department. At the age of 22 von Neumann received his PhD in mathematics from Budapest and a chemical engineering degree from ETH. It was time to choose a career direction. The Göttingen faculty fortuitously intervened, obtaining a Rockefeller Foundation stipend for von Neumann to study with Hilbert.

In 1926 von Neumann settled in Göttingen. At this time physics had its own young rising star. Werner Heisenberg was just two years older than von Neumann. Heisenberg's new theory of quantum mechanics was causing quite a stir, but there was difficulty in reconciling it with a competing formulation of Schrödinger. The young von Neumann made the profound discovery of how to couch Heisenberg's theory in terms of unbounded operators on an infinite dimensional space (known as Hilbert Space). This breakthrough provided a revealing mathematical foundation for quantum mechanics.

Von Neumann went to the University of Berlin in 1927 as a *privatdozent*. He was publishing mathematical research at the stunning rate of one paper per month. This included continued development of his ideas on set theory, quantum mechanics, and operator theory. He also opened investigations into other areas such as game theory. Von Neumann's work came to the attention of Oswald Veblen, who was always on the lookout for new mathematical talent. Veblen and Eisenhart kept an eye on Europe's most promising mathematicians as prospects for the Princeton department. When Weyl resigned from Princeton in 1929, the university was left with a vacant chair and a void in mathematical physics. Veblen wrote Eisenhart urging an appointment for von Neumann:

I am convinced that von Neumann would be a good man to try out. In my last conversation with Weyl, Weyl strongly recommended Neumann as his successor, so he has evidently made good with two of the chief men, Dirac and Weyl. I have had an eye on Neumann for three years or more, as being the best young man in Germany. I met him last summer at Bologna and we talked about the possibility of his coming over on an international fellowship. Later on he wrote me that he would not be able to do it because of having taken a special position of some sort at Hamburg. He is a Privat-Dozent at Berlin, but he is a Hungarian by birth. He makes a very good impression personally.[58]

Von Neumann was invited for a visit beginning February 1930 to give a course on quantum mechanics. The one-term salary was $3,000 plus a $1,000 travel allowance. One feature of the invitation was a query as to whether it might be desirable to invite Wigner to lecture during the same period. Von Neumann was delighted. He was anxious to come to the United States where the opportunities appeared greater than in Europe. Having the companionship of Wigner made it all the more attractive. As it turned out, von Neumann's entourage was growing. He was about to marry Mariette Kovesi.[59]

Arriving in a new country was less intimidating for von Neumann than for most immigrants. He had already lived in Hungary, Germany, and Switzerland. English was among his many languages, and his quick mind made him especially fluent in the discussion of mathematics. When von Neumann landed in New York he found a culture that suited him.

The von Neumann-Wigner arrangement worked out well for the university. The endowment income for the mathematical physics research chair was used to support the two Hungarians each year on a temporary basis. Von Neumann was living half the year in Princeton and the other half in Berlin. His work continued to flourish. He pursued his programs in operator theory and quantum mechanics. The set theory project was abandoned when Gödel's revelations demolished any prospect for success. Among the new directions for von Neumann was ergodic theory, where he and Birkhoff obtained remarkable theorems.

By January 1933 von Neumann was one of the foremost mathematicians in the world. With the continued deterioration in Europe, von Neumann wanted a permanent position in the United States. Eisenhart, having just lost Veblen and Alexander, was working on an offer. This was the situation as Flexner and Eisenhart pondered the contingency over Weyl. When Weyl appeared to accept the Institute position, Eisenhart moved forward on a professorship for von Neumann. Everything changed again when Weyl withdrew, and von Neumann received the fourth Institute professorship.

The substitution of von Neumann for Weyl brought about the third departure of a mathematician from the university faculty. None of these could have occurred without the acquiescence of Eisenhart, who was representing competing constituencies. He was the chief academic officer of the university, the chairman of its mathematics department, and a mathematician. All the while he was operating in a university that was financially strapped by the Depression. Despite the budget shortfall Eisenhart did succeed in keeping Veblen, Lefschetz, Alexander, von Neumann, and Wigner in Fine Hall. If Eisenhart had declined to accommodate Flexner, there would have been ramifications for both institutions. The university lacked the wherewithal to retain all of these scholars. Without Veblen, Alexander, and von Neumann the Institute School of Mathematics might never have been established and brought the Princeton community to its leading profile in the mathematical world.

A Princeton University trustee was likely to have a more territorial perspective. Three of the best university mathematicians had been surrendered. Even balanced with the beneficial addition of Einstein to campus, one of the nation's best mathematics departments had been reduced to a supporting role. Some trustees felt that Eisenhart had acted unwisely. It is difficult to assess whether there was any linkage to subsequent events.

Princeton University was in the midst of a search for the successor to President Hibben. Eisenhart was the choice of the university faculty, but university presidents are chosen by trustees. The Princeton trustees decided on another faculty member. Harold Dodds was the chair of the recently established School of Public Affairs. In May 1933 the trustees selected Dodds as the next university president.[60]

The exodus of Princeton faculty to the Institute ended with von Neumann in January 1933. During the remainder of his tenure as Director, Flexner would make a dozen faculty appointments to three schools. Each of these scholars came from outside the university. There is some evidence that this was a consequence of a secret agreement between the Institute and the university, but the record is sketchy. Both parties had their reasons for maintaining secrecy. What follows is an examination and interpretation of the available evidence.

Under Flexner's successors the faculty received some voice in the selection of new colleagues. There was considerable ambiguity among the faculty as to their understanding of whether Princeton University professors were eligible for consideration. The matter did not come to a head until 1962 when the School of Mathematics considered the appointment of John Milnor from the University.* The Institute Director, Robert Oppenheimer, was operating under the belief that University faculty were out of bounds. He likely obtained this information from Board chairman and charter trustee Samuel Leidesdorf. For, when the mathematicians persisted with Milnor, Oppenheimer solicited the following letter from Leidesdorf:

> It has come to my attention that there has been a discussion regarding the possibility of offering an appointment at the Institute for Advanced Study to one of the professors of Princeton University.
>
> I would like to recall to you a discussion in which I participated with Dr. Flexner and the then President of Princeton University (Dr. Dodds, I believe), which discussion took place many years ago at the time we were establishing the Institute. We pointed out to the President of Princeton the intellectual advantages the Institute would bring to Princeton without in any way interfering with their faculty. The Princeton people were impressed and cooperated with us in every way, even to the extent of offering us the facilities of Fine Hall at a very nominal rental to help the Institute get started. At that time, we agreed that when we undertook to secure faculty members for the Institute, we would never interfere with the Princeton faculty.
>
> We have adhered to this "gentlemen's agreement" through the years and as a result have never had any difficulty with Princeton University. I feel very strongly and know that the Board would agree with me that it would be morally wrong to renege on this understanding now.[61]

Leidesdorf was recalling a meeting thirty years earlier that included Flexner, the university president, and himself. He can be excused for any uncertainty over the name of the president, but there is a fundamental inconsistency in his letter. He states an unequivocal recollection of an embargo and its faithful execution. The problem with this statement comes in fixing a date for the meeting. The entire context of Leidesdorf's second paragraph is set at a time prior to the hiring of the first Institute faculty. However, adherence to the agreement was only possible if the meeting occurred after the von Neumann hiring. When did Leidesdorf's meeting take place?

To date the discussion in Leidesdorf's second paragraph, it is useful to recall that Leidesdorf and Flexner went to Princeton on Saturday, February 27, 1932, to survey possible sites for the Institute. The following day they were

---

*More details are available in [Borel, "The School"] and in Chapter 10.

joined by Maass and Edgar Bamberger. The Princeton excursion occurred just after Flexner's interview with Birkhoff. At this point, Flexner planned to make offers to Birkhoff, Einstein, and Weyl. The hiring of university faculty was totally out of the question. The second paragraph of Leidesdorf's letter fits the vision that Flexner would have pitched to President Hibben on a courtesy visit at this time. The problem with this interpretation is that the subsequent hiring of Veblen, Alexander, and von Neumann would have violated the gentlemen's agreement.[62]

One explanation of the contradiction in Leidesdorf's letter is that he is conflating two meetings in his second and third paragraphs. Stipulating that the above discussion with Hibben took place on the site visit, there is evidence of a later meeting that conforms to Leidesdorf's third paragraph. Beatrice Stern's unpublished history of the Institute describes events taking place after the von Neumann maneuvers:

> Princeton's Trustees and executives were angry, and it became necessary to do something about that. Mr. Bamberger and his close associates ultimately gave a pledge to Acting President Duffield, an old friend of the Founders, that the Institute would take no more men from the University. Naturally, the agreement was secret; only those directly involved knew about it, for the danger to the University was great.[63]

Stern's only reference is a cryptic footnote to a communication from Leidesdorf to Oppenheimer. It is unclear whether her source is the above letter, an interview with Oppenheimer, or something else. In any event she is depicting a different meeting than that in Leidesdorf's second paragraph. It is possible that Stern was taking some liberty in adapting Leidesdorf's letter so that it fit in with the chronology. However, in the 1950s, Stern did an exhaustive search of Institute records and interviewed several of the principle figures. Her account deserves to be taken seriously and the detail makes it seem more likely that she had additional information. If so, it fits nicely into the second of two meetings theory. Following this thread, it is difficult to conclude whether Bamberger's promise was intended to last in perpetuity. Any sensible execution of this intent called for a written record. Leidesdorf was the only participant whose recollections survived, but they are impeached by inconsistency.

Whether or not there was a secret gentleman's agreement, Flexner had no further designs on the Princeton mathematics department. He was truly content with his School of Mathematics faculty. Einstein, Veblen, Alexander, and von Neumann were genuine world-class scholars. Moreover, their nationalities were balanced, and their age distribution was ideal. The roster would not surpass Göttingen, which still had Weyl, but Flexner's negotiation experiences had caused him to moderate his goals.

There was also the worsening economic situation to consider. Early in 1933 the United States banking system collapsed. Franklin Roosevelt took office in March. Flexner was skeptical of the efficacy of Roosevelt's economic programs. Inflation was a major concern, especially with the Institute endowment fixed for the foreseeable future. If the Institute were to be more than a school of mathematics, it was time to put the evaporating resources into other programs.[64]

Veblen had his own ideas. From the start he had his eye on the brilliant British physicist Paul Dirac. Dirac, who was just 31, would win the Nobel Prize later that year. Veblen argued that Dirac could be a scintillating collaborator for Einstein as well as a good fit for himself and von Neumann.[65]

Flexner responded that expansion of the mathematical group was impossible under the current economic situation. Flexner was also troubled by the notion of hiring another foreigner. It was an issue that he had grappled with for years. Now he decided to confide in Veblen.

> Mr. Bamberger and Mrs. Fuld were very anxious from the outset that no distinction should be made as respects race, religion, nationality, etc., and of course I am in thorough sympathy with their point of view, but on the other hand if we do not develop America, who is going to do it, and the question arises how much we ought to do for others and how much to make sure that civilization in America advances.[66]

Veblen was sympathetic. He amended his proposal to a position for the American Norbert Wiener and a one-term visit for Dirac. The landscape, however, was about to change drastically. The Flexner-Veblen discussion took place late in March, just before the Nazis cleansed the Jews from the German faculty.[67]

Hitler began consolidating his grip on power immediately after his appointment as chancellor on January 30. One day later he succeeded in having the Reichstag dissolved. Hitler was counting on the March 5, 1933, election to strengthen the Nazi representation. There were gains, but less than he anticipated. Even so, it was sufficient for him to maneuver the new Reichstag into surrendering its authority on March 23. With this "Enabling Act" Hitler took a major step toward gaining dictatorial power. Meanwhile an anti-Semitic campaign was underway. Sporadic harassment and beatings of Jews were occurring in the streets. On April 1 there was a boycott of Jewish businesses. Then on April 7 came the "Law for the Restoration of the Professional Civil Service." Individuals of non-Aryan descent were removed from jobs such as university professorships.

Einstein and Weyl were among the millions of people affected by these developments. For example, the April 7 law meant that neither Einstein nor Weyl's children were eligible for civil service positions. Weyl, who for the time being could hold a professorship, was preoccupied with other problems. He had been unable to function following his January 11 telegram to Flexner. Sunk in a deep depression and unable to sleep, Weyl could neither teach nor do research. At the urging of his colleagues Weyl entered a sanitorium in Berlin. There he received treatment while in virtual isolation from the outside events.[68]

In early February Weyl succeeded in having a letter smuggled out to Veblen. It provided a poignant picture of Weyl's misery and hopelessness. He saw no prospect of ever again doing mathematics or returning to Göttingen. A letter to Flexner at the end of the month indicated some improvement. Weyl confessed that his self-confidence was shattered, but he saw himself in the middle of recovery from a serious illness. He begged Flexner to forgive the trouble that he had caused. Flexner and Veblen responded with supportive letters.[69]

Weyl was released from supervision in March. For some time he had been insulated from the startling changes occurring in Germany. Rather than return immediately to Göttingen, Weyl went to the Swiss resort Lugano to speed his convalescence. There he learned of the Reichstag developments, but, away from Germany, it was impossible for him to grasp the social deterioration taking place in his native country. At the end of the month Weyl wrote Veblen that he was feeling much better and had regained some of his confidence. In the long letter Weyl spoke of his intention to resume mathematical research and teaching. Weyl explained that he would soon return to Göttingen, but was pessimistic about his future in Germany. Then Weyl began to explore whether he had really burned his bridge to Princeton. Weyl confided that he now realized that the Institute was the best place for him. While he understood that it was not possible to restore the prior offer, Weyl asked if he might visit the Institute and be given a couple of years to prove himself.[70]

In mid-April Weyl went from Lugano to Zurich. There he came into contact with mathematicians for the first time since his institutionalization. As Weyl learned of the ongoing developments in Germany, he faced a dilemma of whether or not to return to Göttingen. His savings were exhausted and there was an urgent need to earn support for his family. The only immediate guaranteed income was at Göttingen, but Germany was a dangerous place for a household with a Jewish wife. If Weyl returned to Göttingen he feared that he would never be permitted to leave.[71]

Veblen wanted to help Weyl, as well as the other mathematicians who were being victimized. Flexner was horrified by what was occurring in Germany to the educational system he once cherished. He held no grudge over Weyl's past behavior and was willing to consider him for a fifth professorship in mathematics. The real concern was whether Weyl was fully recovered from his mental crisis. Flexner decided that he and Veblen needed to interview Weyl and assess his condition. To bring Weyl to America, Flexner worked with Aydelotte, the president of Swarthmore. Weyl's illness had forced him to cancel a series of lectures at Swarthmore. Aydelotte arranged for renewal of the invitation. The plan was for Weyl to deliver the lectures in the fall.[72]

Weyl remained in Zurich as long as possible, monitoring the worsening conditions in Germany and communicating with Veblen. In late April, just before leaving for Göttingen, Weyl sent an update. He hoped that his German stay would be brief, just until he could arrange for another job. It would be difficult to conduct the job search from Göttingen. Letters could only go out via trusted friends who were passing through Germany. He would seek permission to visit Swarthmore, but circumstances might force an earlier departure. Weyl confided to Veblen that he was desperate and amenable to a lesser position than he had turned down at the Institute. There was some prospect in Madrid.[73]

Back at Göttingen, Weyl joined a faculty that had been decimated by the purge of Jews from the civil service. It was even worse than he had anticipated. The original law had carried an exception for veterans of World War I. This seemed to spare mathematics professor Richard Courant, a close friend of Weyl who headed the mathematics institute. Weyl reached Göttingen as Courant and others were dismissed from their duties.[74]

Weyl was forced to assume Courant's position as director of the mathematics program. Weyl, with no taste for administration, had some hopes of bringing back Courant and other dismissed Jews. After a month, Weyl believed that he was making progress. This unfounded optimism lent a somewhat upbeat perspective to a letter carried out by von Neumann:

> It has calmed down here a little; for a good part of the university is still under the despotism of the radical student body and the sword-swaggering

national-socialist Privatdozenten. One can only communicate with them in a particular jargon. The government only interferes in certain situations. As department head of the Institute I have to howl a little with the wolves. But I did not have to struggle yet with an inner conflict in me. My position seems secure; I don't know why this is. I can take some risk in controlling the nonsense.[75]

Weyl was deluding himself. Compare this historical view of the period:

During the spring and summer of 1933, Germany fell into line behind its new rulers. Hardly any spheres of organized activity, political or social, were left untouched by the process of *Gleichschaltung*—the 'coordination' of institutions and organizations now brought under Nazi control. Pressure, from below, from Nazi activists, played a major role in forcing the pace of the 'coordination'. But many organizations showed themselves only too willing to anticipate the process and to 'coordinate' themselves in accordance with the expectations of the new era. By the autumn, the Nazi dictatorship—and Hitler's own hand—had been enormously strengthened.[76]

There was never any chance for Courant. Weyl was completely overmatched on this and everything else. On July 10 Veblen received the following cable from Weyl: "Situation so threatening that I should accept visiting professorship Madrid December to April with regular leave from Göttingen if possible. Cable advice to Scherer."[77] Veblen's response was to ask Weyl to delay his decision on Madrid. With Flexner at his isolated Canadian retreat, Veblen needed time to work for a counteroffer.

Flexner made his own analysis of the situation. There was no doubt that Weyl would accept an offer, giving the Institute the world's premier mathematics group. The medical questions had been addressed by one of Weyl's doctors who had recently passed through New York. Flexner decided that it was time to act. The only remaining obstacle was a possible Bamberger veto. One year earlier Bamberger had balked at enhancing Weyl's offer. It was only following a rare Flexner appeal that Bamberger had agreed to go along. The subsequent events had convinced Bamberger that his original instincts on Weyl were correct.[78]

Flexner wrote to Bamberger, seeking approval of an offer to Weyl. Once again Bamberger expressed his opposition. Flexner decided that, on Weyl, the ground was too shaky to ask for an immediate reconsideration. If Bamberger held firm then the game was over. Instead, Flexner merely asked Bamberger to withhold judgment until they had an opportunity to meet in person. Flexner included a review of Weyl's case, presented in the context of building a school of mathematics. In this account Weyl's actions were portrayed honorably when contrasted with those of Birkhoff.[79]

Flexner, dreading the prospect of confronting Bamberger, needed a surrogate. The exchange with Bamberger took place as Aydelotte was arriving for a vacation at Flexner's Canadian home. The timing was perfect. Flexner knew that Bamberger was impressed by Aydelotte. Over the past six months Flexner had taken Aydelotte into his confidence on Weyl, Birkhoff, and a number of other Institute matters. Flexner had come to think of Aydelotte as his "understudy" and heir-apparent as Director of the Institute. Aydelotte had acted on Weyl's behalf in Germany the previous summer and was expecting to host the Swarthmore visit in the fall. Flexner updated Aydelotte on the latest Weyl developments.[80]

With Aydelotte slated to be his successor, it was appropriate for Flexner to arrange a meeting where the Aydelottes and the founders had an opportunity to become better acquainted. This took place in the middle of August on the Jersey shore. The topic of Weyl arose quite naturally. Bamberger was a good listener who was interested in obtaining the views of Aydelotte. By the end of the conversation Bamberger had withdrawn his objections to Weyl. The way had been cleared for Weyl, with only the pro forma approval of the Executive Committee to be obtained.[81]

The $15,000 offer went out on September 7 in care of an intermediary in Zurich. Everything was kept very secret so as not to jeopardize Weyl's chances of leaving Europe with his family. In October the Weyls succeeded in crossing the ocean. They reached the United States just one week after the Institute's other German refugee. Einstein's path to the Institute provided entirely different challenges. These Flexner handled with less equanimity.[82]

In mid-March 1933 Einstein completed his winter residence at Caltech. The two-week return trip to Europe coincided with the escalation of anti-Semitism in Germany. As a prominent German Jew, Einstein became a target of the campaign. He was attacked in German newspapers. Both of his residences were ransacked.[83]

Einstein's reaction was to assert himself in solidarity with his Jewish countrymen. He became the champion of the German Jew. Einstein's outspoken positions were vilified in Germany. When Einstein reached Europe it was impossible for him to return to Germany. He resigned his German scientific affiliations and took up temporary residence in Belgium. Einstein's refugee status became a major item in the world press. Universities reacted by offering professorships to show their support.

Flexner was disturbed by newspaper reports that Einstein had accepted positions in Madrid, Brussels, and at the Collège de France. The Institute was paying a full-time salary and expected exclusive custody. Flexner wrote Einstein:

> Of course, this does not mean that professors are not free in their vacations or sometimes during the year to deliver a scientific lecture or several lectures at some other institution, but, if the title of professor elsewhere is accepted, the Institute will be regarded, not as a place which will protect men from distraction but as one of several places in which their activities may be carried on.[84]

Einstein held a different view of his obligations to the Institute. He had agreed to split his time between Princeton and Berlin. Since a return to Germany had become impossible, he felt free to reallocate those six months for vacation and new obligations. The offers from the University of Madrid and Collège de France arrived through diplomatic channels. The positions were largely symbolic, representing support for Jewish human rights. Einstein felt an obligation to accept. He was unable to see how his holding an additional title caused any injury to the Institute. Irreconcilable differences were emerging. Einstein could never understand the marketing value of his name. Flexner thought that Einstein should stay out of the limelight and remain silent on political issues.[85]

One aspect of the Madrid offer reopened the sensitive area of the earlier Einstein-Flexner negotiations. Madrid was willing to add on Mayer as a professor. Einstein, still disappointed over Flexner's treatment of Mayer, requested that the offer be matched and Mayer's Institute salary increased. Flexner faced a difficult decision. If the accommodation were refused there was the real possibility that the Institute would lose Einstein. Even if it were not Madrid, Einstein was such a magnet for job offers that he could probably leverage a position for Mayer elsewhere.[86]

There was a great deal at stake, but Flexner could not bear the thought of making Mayer a professor. He decided to hold firm and stand up to Einstein. Flexner pointed out that junior collaborators were rarely appreciated when they remained in the shadow of their mentor. He urged Einstein to make the sacrifice of allowing Mayer to go out on his own to Madrid where he could become recognized as an independent scholar. Subsequent developments in Spain made it a less congenial place for Einstein. Neither Mayer nor Einstein would join the Madrid faculty. Flexner's gamble had paid off, but their relationship had soured. In a late July letter, Einstein asked Flexner whether he wished to terminate their agreement.[87]

During the summer of 1933 Einstein traveled between Belgium and England, working on science and doing what he could to help Jewish refugees. The latter involved contributing his name and, occasionally, his presence. Among his commitments was an October 3 fundraiser in London's Albert Hall. Making this appearance and speech would delay his arrival in Princeton until the middle of October. The timing provoked more unpleasantness.

When Flexner and Einstein met one year earlier in Caputh, it was unclear precisely when the Institute would open. Flexner's original offer to Einstein described the academic calendar as follows: "Term shall run from approximately the beginning of October to approximately the middle of April."[88] The 1933 opening of the Institute was subsequently scheduled for Monday, October 2. When Flexner learned that Einstein planned to arrive in mid-October, his response took on an uncharacteristic tone of bitterness and condescension. Aside from Flexner's frustration with Einstein, there were other forces that may have caused him to abandon diplomacy. The 66-year-old Flexner had had a tonsillectomy in June. Over the summer he was suffering from complications of the surgery. It is also understandable that he wanted his star, and only non-Princetonian, faculty member in residence for the Institute opening. Still, it made no sense to jeopardize his relationship with Einstein over this issue. Although Flexner's tactics failed to secure Einstein for the opening, they did foreshadow his designs on control.[89]

Flexner was sympathetic to the causes that Einstein was leading. He just believed that Einstein's political activities were ill-advised and dangerous. Flexner was informed that Nazi groups were operating in the United States. If Einstein made speeches and appearances he was inviting an assassination attempt. The only protection, according to Flexner, was for Einstein to be secluded in Princeton, devoting himself entirely to his research. Flexner was convinced that Einstein was too naive, and his wife too publicity hungry, to appreciate the danger they faced. Flexner took up the task of insulating Einstein, justifying his actions from their conversation in Caputh. At that time the Einsteins mentioned that they would be grateful for Flexner's protection from publicity. Flexner took this innocuous comment as conferring a blanket authorization to act on Einstein's behalf.[90]

The program went into effect when Einstein reached the American shore. Flexner knew that Einstein's previous landings in the United States had aroused a carnival atmosphere. Working secretly with the State Department,

Flexner made arrangements for Maass to meet the Einstein party as their ship left quarantine. Maass took them in a small boat to the New Jersey side where cars were waiting for the trip to Princeton. The local New York authorities were unaware of these arrangements. When the Einsteins' ship docked in Manhattan, the Mayor was waiting in the rain to lead a celebration. The politicians were humiliated to learn that Einstein had bypassed Manhattan and was already in New Jersey.[91]

With Einstein ensconced in Fine Hall, Flexner was in a position to screen mail and phone calls. Based on the Caputh understanding, Flexner took the liberty of rerouting Einstein's mail through his own Princeton office. Einstein was unaware of the invitations that were declined on his behalf. Two weeks after Einstein's arrival in America, Flexner's abuse reached absurdity when he responded with these regrets to President Franklin Roosevelt:

> Dear Mr. President:
>
> With genuine and profound reluctance, I felt myself compelled this after-noon to explain to your secretary, Mr. MacIntyre, that Professor Einstein had come to Princeton for the purpose of carrying on his scientific work in seclusion and that it was absolutely impossible to make any exception which would inevitably bring him into public notice.
>
> You are aware of the fact that there exists in New York an irresponsible group of Nazis. In addition, if the newspapers had access to him or if he accepted a single engagement or invitation that could possibly become public, it would be practically impossible for him to remain in the post which he has accepted in this Institute or in America at all. With his consent and at his desire I have declined in his behalf invitations from high officials and from scientific societies in whose work he is really interested.
>
> I hope that you and your wife will appreciate the fact that in making this explanation to your secretary I do not forget that you are entitled to a degree of consideration wholly beyond anything that could be claimed or asked by any one else, but I am convinced that, unless Professor Einstein inflexibly adheres to the regime which we have with the utmost difficulty established during the last two weeks, his position will be an impossible one.
>
> With great respect and very deep regret, I am[92]

Flexner did not succeed in stopping all communications. Another admin-istration official did reach Einstein and mentioned the declination. Einstein was shocked to learn how he was being handled. His apology to the President followed and led to a dinner in the White House early the following year.[93]

In Einstein's first month at the Institute there were a number of unpleasant exchanges with Flexner. When Einstein requested a stipend for a promising young physicist, Flexner did not just refuse. He pointed out that by his late arrival Einstein had forfeited his voice in such decisions. The dispute over Mayer continued and there was a major battle over whether Einstein could play his violin in a benefit concert for some refugee friends.[94]

Flexner candidly expressed his view of the Einsteins to Maass:

> I am beginning to weary a little of this daily necessity of "sitting down" upon Einstein and his wife. They do not know America. They are the merest children, and they are extremely difficult to advise and control. You have no idea the barrage of publicity I have intercepted. I should suppose that half of my time is devoted to protecting Einstein. It will be worth while if I succeed in doing it permanently within this year.[95]

When Einstein protested to Flexner over his meddling, Flexner reluctantly agreed to cease his interventions, but it did not change his opinion.[96]

The relationship between Flexner and Einstein was off to a horrible start. Even with the mail properly routed, other matters came to Flexner's attention. A commercial dealer delivered a piano to Einstein as an unsolicited gift. Einstein acknowledged the present, using Institute letterhead. Without Einstein's permission, his letter was placed in the store window. Flexner protested the exploitation to the merchant. The conversation naturally included some discussion of Einstein. Out of this dispute and others, Einstein heard from third parties that he was being impugned by Flexner. An insulted Einstein began declining social invitations from Flexner and threatened to withdraw from the activities of the Institute.[97]

Word of the feud reached Bamberger, who deplored relations of the sort. When Flexner recognized that a reconciliation was essential, he turned on the charm in a long December meeting with Einstein. The two men made up and put the past behind them. The impasse over Mayer's status evaporated with the soon-ended Mayer-Einstein collaboration.[98]

The row between Flexner and Einstein was the only blemish on the Institute's phenomenal first year of operation. From the start Flexner had said that the success of the undertaking depended on whether he could engage first-class scholars. The faculty in the School of Mathematics more than fulfilled this ambition. It was in the educational component, however, that Flexner aspired to establish a new paradigm. This indeed did happen, but it was Veblen's vision that shaped the final outcome.

In the fall of 1932 Flexner amended his agenda from graduate to post-doctoral education. Flexner's intention was for the Institute to provide its students with a transitional research period between the completion of the PhD and assumption of a first faculty appointment. Institute professors would mentor new PhDs to the point where they were better able to pursue independent programs. These students would then be prepared to begin faculty positions at American colleges while maintaining their scholarship. Veblen took a broader view. It was not just the fresh PhD who would benefit from the Institute environment. For faculty who were early in their careers, a year at the Institute could stimulate their research at a crucial time. Veblen saw the Institute community including scholars in varied career stages, emphasizing the earlier ones.

Veblen sought to arrange one-year visits from promising young mathematicians who already held positions at other universities. The difficulty was that junior faculty tended to be ineligible for sabbaticals or other types of support. Veblen proposed that, in some cases, the Institute provide stipends so as to share the salary burden with the visitor's home institution. Flexner was at first opposed to diverting Institute resources for the benefit of other universities' faculty. Veblen replied that the Institute faculty would be enriched by their interaction with this group. In January 1933 Veblen persuaded Flexner that the concept was worthy of experiment. Among the visitors in the Institute's first year were Adrian Albert from the University of Chicago and Egbertus van Kampen from Johns Hopkins.[99]

Altogether there were 23 postdoctoral "workers" during the 1933–1934 academic year. These included Kurt Gödel, four Americans on National Research Fellowships, and three Europeans on Rockefeller Foundation Fellowships. Several were already established as scholars. Only a few were new PhDs. The group connected the faculty to the Princeton graduate students in a manner that may not have occurred under Flexner's polar faculty-student scheme. It all contributed in a magical way to create a vibrant mathematical atmosphere. When Flexner visited Fine Hall from his downtown office, he must have been reminded of his Johns Hopkins days. Research was indeed in the air.

Veblen's successful visitor scheme was quickly embraced by Flexner. A sum of $20,000 was budgeted for grant-in-aids the following year. Johns Hopkins was so satisfied that it provided full support for its associate professor Oscar Zariski in 1934–1935. The postdoctoral roster grew to 35 that year. The program remains in effect today on a similar basis. It is one of the distinguishing features of the Institute. In the early years the term "worker" was replaced by "member."[100]

The worker/member gave the Institute a scholarly group that compared in some way to the *privatdozenten* in Germany. The Institute faculty and members together with the university faculty and graduate students made Fine Hall into a mathematical research community that was unprecedented in America. It was a coincidence that the Institute opened in 1933 just months after the Nazi programs removed so many able faculty and *privatdozenten* from Göttingen. The net effect was an abrupt transoceanic shift in the world's most prestigious locale for mathematical research. Princeton replaced Göttingen.

It would never again be necessary for Flexner to court a mathematician. The Institute's first year of operation was so successful that scientific superstars began making inquiries about possible positions. In the summer of 1934 both Erwin Schrödinger and Marston Morse expressed their interest. Schrödinger had shared the 1933 physics Nobel Prize with Dirac. Morse was a Harvard mathematics professor who received the American Mathematical Society Bôcher Prize in the same year. Both Schrödinger and Morse were in their forties.[101]

The Austrian-born Schrödinger was a former colleague of both Weyl and Einstein. In 1921 Schrödinger became a professor of theoretical physics at ETH. In Zurich he had substantial discussions with Weyl on mathematical physics. There were more personal connections as well. Erwin and Anny Schrödinger had an open marriage. Weyl and Anny began an affair.[102]

In 1926 Schrödinger formulated the famous partial differential equation for wave mechanics that bears his name. Following a decade at ETH Schrödinger accepted a prestigious chair in Berlin. There, he and Einstein found each other to be valuable sounding boards for ideas. Their interaction was halted by the mass exodus from German universities in 1933. Many brilliant scholars suddenly were unemployed. Schrödinger obtained a temporary position at Oxford.

During this time of worldwide depression, professorships were scarce. One opportunity was at Princeton where the chair of mathematical physics remained vacant. Eisenhart was always on the lookout for European talent, but cautious in making commitments. He invited Schrödinger to Princeton as a visiting lecturer for early 1934. It was an audition of sorts for a more permanent arrangement. Schrödinger arrived in March and was reunited in Fine Hall with Einstein and Weyl. One month later Schrödinger returned to England, holding an offer for the university chair in mathematical physics.[103]

Princeton was on the verge of achieving world supremacy in theoretical physics. Dirac had already agreed to a visiting Institute professorship for the following year. If Schrödinger came to the university, then both current Nobel physics laureates would be joining Einstein, von Neumann, Weyl, and Wigner in Fine Hall. Of special significance to Eisenhart and Flexner was that, for the first time in their collaboration, the Institute was cast in the supporting role of a major recruitment. It was an important step in Institute-university relations.

Einstein, Weyl, and Veblen each worked to overcome Schrödinger's qualms over moving to America. The results were mixed. Schrödinger declined the university offer, explaining that he was expecting more attractive terms from the Institute. It was a huge embarrassment for Flexner. Flexner quickly replied to Schrödinger, disabusing him of any prospects at the Institute and urging the resumption of negotiations with the university. The letter was carboned to Eisenhart along with an awkward apology and denial of Institute sabotage. Schrödinger remained in Europe.[104]

There was a sharp contrast in Flexner's reaction to an overture that arrived just two months prior to Schrödinger's. Veblen learned then of Marston Morse's interest in an Institute professorship. For several years Veblen, Alexander, and Lefschetz had coveted Morse for the Princeton mathematics department, but the university administration declined to sanction another senior mathematics line. Morse's name had arisen frequently in Flexner's consultations on mathematics. Morse was regarded as the best remaining American prospect. His sudden interest in joining the School of Mathematics was welcomed by Flexner. There was no other mathematician in the world whom Flexner would have given consideration for a sixth professorship.[105]

Harold Calvin Marston Morse was born in Waterville, Maine, on March 24, 1892. Marston was his mother's maiden name. Ancestors on both sides had lived in the United States since the 1630s. Morse received his early education in Waterville, including an undergraduate degree from Colby College.[106]

In 1914 Morse entered the graduate mathematics program at Harvard. Three years later he completed his PhD under the direction of G. D. Birkhoff. Morse's thesis involved geodesics on a surface of negative curvature. He characterized the geodesics in terms of the generators for the fundamental group of the surface. Graduation coincided with the United States' entry in World War I. Morse enlisted in the army and saw action in Europe.

Returning from the War in 1919, Morse began an academic career in mathematics. Early postings took him on a tour of the Ivy League from Harvard to Cornell to Brown and back to Harvard in 1926. While at Cornell

Morse married Celeste Phelps. He also produced his first paper on a subject that was to become known as Morse Theory.

Morse Theory is one of the seminal mathematical developments of the twentieth century. Its essence is that there are fundamental relationships between calculus and algebraic topology. These emerge for smooth real valued functions defined on a manifold. A basic object of interest in calculus is the critical points of such functions. Most functions have a finite number of these points and they can be classified by their second derivative. In one variable calculus the distinction is between maxima and minima. Thus in dimension one there are $2 = 1 + 1$ types of critical points. For functions of two variables there are maxima, minima, and saddles (i.e., $3 = 2 + 1$ types). With a manifold of dimension n there are n variables and $n + 1$ types of critical points. The type numbers of the function are the $n + 1$ numbers specifying the quantity of each type of critical point. The question is whether the nature of the manifold places any inherent restrictions on the type numbers that can be realized by a function on the manifold.

The theory of algebraic topology associates $n + 1$ nonnegative numbers to each manifold, called Betti numbers. Morse showed that there are $n + 1$ linear inequalities that systematically relate the Betti numbers to the type numbers (actually, one of these is an equality). Thus the algebraic topology of the manifold establishes criteria that must be met by the critical points of a typical smooth function on the manifold. This is one special manifestation of Morse Theory. Morse and others went on to develop a far reaching understanding of these principles. In the second half of the twentieth century, Morse Theory played a vital role in several major mathematical discoveries.

The mathematical community rapidly perceived the significance of Morse's discovery. In 1931 Morse was invited to deliver the prestigious Colloquium Lectures to the American Mathematical Society. It was in the same year that Flexner began accumulating nominations for his faculty. Morse was among the geometers suggested by Lefschetz. With Flexner seeking a single mathematician, Morse was overshadowed by his Harvard mentor Birkhoff. In 1932 Birkhoff withdrew. When Veblen and Weyl pushed for a larger mathematics group, Morse's name emerged at the top. Weyl ranked Morse as the top American, ahead of Lefschetz and Wiener.[107]

Veblen then informally approached Morse about the possibility of coming to the Institute. Morse was not interested at the time, but Veblen asked to be apprised if his feelings were to change. As the Institute flourished over its first year, Morse kept the opportunity in mind. Harvard, with Birkhoff and its strong graduate students, remained an ideal professional location for Morse. An awkward personal situation was that Morse's ex-wife had married his mathematics colleague Osgood.

Morse redeemed his rain check in 1934. Flexner was then actively seeking faculty for other schools. Despite the ongoing economic uncertainty and the already large commitment to mathematics, Flexner was pleased at the prospect of adding a first-class American mathematician. The biggest obstruction was the delicacy of relations with Harvard. Flexner's old adversary, President Lowell, had recently been succeeded by James Conant. If the Morse matter were handled in a manner that was courteous to Harvard authorities, there was the prospect of getting beyond the hard feelings of the past. Flexner proceeded deliberately and made the offer in October. The salary was set at $12,500, midway between the other levels in the school.[108]

Marston Morse joined the School of Mathematics in 1935, completing its first generation of faculty. Einstein, Veblen, Alexander, von Neumann, Weyl, and Morse were an extraordinary group. Each year they were joined by visitors with diverse interests and backgrounds. In 1934–1935 (the second year of operation) these included Paul Dirac, Oscar Zariski, Richard Brauer, Georges Lemaître, Carl Siegel, Jesse Douglas, Deane Montgomery, Joseph Walsh, and Alonzo Church. The School of Mathematics was well established.

Felix Frankfurter outside 192 Brattle Street, ca. 1938. (Courtesy of Art & Visual Materials, Special Collections Department, Harvard Law School Library.)

- CHAPTER SEVEN -

LAUNCHING THE OTHER SCHOOLS

With the success of mathematics, Flexner's challenge was to create other schools of comparable distinction. Next in line was economics, and the stakes were high. Bamberger questioned whether economic research was of any practical use. His skepticism was seen by Flexner as an opportunity to demonstrate the value of the Institute. If the School of Economics were to tangibly advance the American system, then Bamberger would surely entrust the remainder of his fortune to the Institute.[1]

One of Flexner's earliest objectives was to identify a person to lead the Institute program in economics. Flexner asked prominent scholars to recommend the best mid-career American economist. Out of these discussions there emerged an overwhelming consensus for Jacob Viner of the University of Chicago. However, Viner failed to impress Flexner in their 1931 meeting in Paris. Viner remained under consideration, while Flexner continued his search.[2]

The next front-runner for the economics/history position was one of Flexner's consultants. Flexner held Charles Beard's scholarship in high esteem. They became friends during their continuing conversation over plans for the Institute. Their views were compatible on many of the important issues. When a difference of opinion arose, civility was always maintained. A pattern emerged of breaking down the controversy so as to identify areas of agreement and then joking about the remaining differences. Flexner thrived on the interaction. He relished the thought of working with Beard as a member of his faculty.

Early in 1932 economics moved to the back burner as attention and resources were devoted to recruiting mathematicians. When Flexner returned to economics at the end of the year, the 58-year-old Beard was no longer under consideration. Age was one factor in Beard's elimination. It is also possible that Beard was too controversial for the conservative Bamberger.

159

At this point Flexner had the benefit of his own experience in creating the School of Mathematics. If economics were to follow the mathematics template, then it was essential to secure a professor such as Veblen. Veblen was not the greatest mathematician in the world, but his judgment and connections were indispensable in planning and executing the School of Mathematics. By late 1932 Flexner had come to rely on Felix Frankfurter as he had on Veblen one year earlier. In November Flexner closed a letter to Frankfurter with a feeler about joining the faculty. When there was no immediate response to the overture, Flexner's optimistic reaction was that it had not been rejected.[3]

Frankfurter may have seemed an odd choice to head a School of Economics. After all, he was a law professor, not an economist. Yet through his research, teaching, and service Frankfurter cultivated an impressive network of connections in the academy and government. His interests covered a broad spectrum that included politics, society, labor, and history. Frankfurter's range and insight were suited to serving Flexner's vision for the School of Economics.[4]

During the Institute's conceptual stage, Flexner set a mission of eliminating panics and depressions from the business cycle. In his view these recurring economic crises were just one consequence of a broader problem. Flexner believed that existing economic theory was inadequate. Economics could no longer be viewed "as a separate science concerned with exchange or transportation or profit or loss but as one factor in the organization of society."[5] To understand economics it was essential to account for the contribution of political and social forces. Flexner began seeking to assemble a multidisciplinary group "to work at different aspects of the problem in entire independence and yet in the hope and expectation that their results would bear upon one another."[6] To oversee and integrate these diverse efforts required a scholar with the wide-ranging expertise of Frankfurter.

Flexner's design for the economics school was brushing up against some of the Institute's founding principles. Every professor was to have complete liberty in pursuing a research program. For example, a mathematician selected under the rubric of geometry was free to work independently in algebra or analysis. In the second school research was to be, to some degree, both targeted and collaborative.

Flexner confronted an even more sacred tenet in recruiting economics personnel: "I assume at the outset that no subject will be chosen or continued unless the right man or men can be found." Frankfurter's participation was at best questionable. The prospects for economists were even worse. Viner, who continued to elicit raves, was the only economist under serious consideration. Flexner had in mind a political scientist and an historian for the project, but the overseer and economists were essential to the viability of the school.

Flexner had demonstrated his patience in the recruitment of Weyl. He could wait a year or two for Frankfurter to come around and a brilliant young economist to emerge. Until these crucial personnel issues were resolved, it was sensible to delay the opening of the school and the hiring of other faculty. The timing became more complicated when another opportunity arose for David Mitrany, the political scientist whom Flexner was counting on for the economics team. If Mitrany were not immediately hired by the Institute, then it was possible that he would be lost forever.[7]

David Mitrany was born January 1, 1888, in Bucharest, Rumania. He began his undergraduate study in Hamburg and at the Sorbonne, going on to the London School of Economics where he received BSc, PhD, and DSc degrees. During World War I Mitrany was stationed in London, representing Rumania in the Intelligence Department of the British Foreign Office. With this background he was fluent in several languages.

Mitrany's expertise in foreign affairs brought him a variety of editorial work and consultantships. His resumé included associations with the *Manchester Guardian*, the Carnegie Endowment, and the British Labor Party. In 1931 Mitrany began a two-year appointment as visiting professor of government at Harvard.

Flexner had known Mitrany and his wife Ena for several years. When Flexner's daughter studied at Oxford, the Mitranys were among the network of friends she inherited from her father. The families remained in touch. Abe kept David apprised of developments with the Institute, sending him report drafts for criticism. Flexner's motivation went beyond mere vetting. The Institute would need an able social scientist with a combination of academic and real-world experience. From time to time Flexner mentioned to Mitrany the possibility of his being called to the Institute.[8]

In January, 1933 Mitrany had one semester remaining on his appointment at Harvard. In seeking his next position, Mitrany's flexibility was limited by the health of his wife. The best opportunities were in America, but Ena was in England undergoing extended therapy for a psychiatric illness. Mitrany was hoping for an arrangement permitting him to spend time with Ena while she were cured, and then to move as a couple to the United States.[9]

Mitrany outlined for Flexner a long-term project in political science. It was directed at the "International Implications of National Economic Planning." He would undertake "a study of Politics from a purely sociological angle, untrammeled by any a priori association with existing political divisions and institutions or by any impulse to provide argument for or against them."[10] Mitrany proposed to begin with a year of preliminary study, "on my own resources." He was asking that the Institute commit to picking up support at the end of the first year.

Flexner was immediately positive. Both Mitrany and the project seemed ideal for the economics school. Even the year in England fit nicely into the timetable, allowing time to identify the other personnel. Flexner reiterated his intention to include Mitrany in the economics program. A few days later Flexner learned that Mitrany was under consideration for a position at Yale. When Harvard began expressing interest, Flexner decided it was time to act.[11]

An important step was to obtain an evaluation by Frankfurter. Frankfurter was enthusiastic about Mitrany, noting his "scientific temper" (unusual in politics), "eye for essentials," and "universal outlook" toward the social sciences.[12] Flexner went ahead with the offer to Mitrany, but he was not yet ready to commit to an entire school. There would be no public announcement of Mitrany's appointment. His salary was set at $6,000 while on leave abroad, and $10,000 when he began residence at the Institute. Mitrany had the security to support Ena and pursue his work. With this stealth maneuver Flexner could still abandon what he began referring to as the School of Economics and Politics.[13]

Meanwhile Flexner embarked on a convoluted scheme to recruit a professor who would give the School the sort of leadership that Veblen had provided for Mathematics. The retirements of Bamberger and Fuld created two vacancies on the Board of Trustees. In filling these positions it was desirable to select social science experts who would assist with the formation of the new school. The hidden agenda was that a Board member, already invested in the Institute, might be more amenable to joining the faculty. Flexner nominated Frankfurter and Walter Stewart, marking the emergence of Stewart as a significant Institute figure.

In 1885 Walter Stewart was born in Manhattan, Kansas. From his farm upbringing Stewart went on to study at the University of Missouri. Next he pursued an academic career, teaching economics at Missouri, Michigan, and Amherst. In the mid-1920s Stewart left the academy and took on jobs in the banking and securities business. He became associated with the Federal Reserve and Wall Street investment firms.

Stewart was becoming an influential figure behind the scenes of the American economic system. His reach became international when Stewart served a stint as economic advisor to the Bank of England. A consummate insider, Stewart participated in world banking decisions while keeping a low profile in the press. It was through British connections that Stewart came to Flexner's attention. By this time Stewart was back in America, leading a Wall Street investment house.

There was already a significant link between Stewart and Flexner. Stewart was among the newer generation who were installed on the Rockefeller charity boards after Flexner's withdrawal. This common interest was the likely starting point for their two-hour lunch conversation in March, 1932. The main topic of discussion, however, was the state of economic research in American universities. Stewart, like Frankfurter, agreed with Flexner on its deplorable condition. While this left quite a bit of latitude as to how the Institute might provide a remedy, Stewart had made a most favorable impression.[14]

In February 1933, Frankfurter and Stewart accepted positions on the Institute Board. Their terms would begin later in the year. Both men were positioned to exert considerable influence either from the Board or, if they chose, on the faculty. A striking difference between the two was in their manner of delivering advice. With or without solicitation, Frankfurter could be counted on to record a forthright and unambiguous opinion. His written assessments of Viner, Mitrany, and Schumpeter leave no doubts.[15]

In contrast, there are few records of Stewart's views. Stewart preferred to offer his advice at private luncheons. Even when Flexner requested an opinion at a trustee meeting, Stewart insisted that it be excluded from the minutes. On occasion Flexner conveyed these remarks to other individuals. The letters provide some flavor of Stewart's positions, but could be distorted by the spin that Flexner intended for the recipient.[16]

It would be some time before there was significant movement on economics. For the remainder of 1933 Flexner was occupied with the School of Mathematics, particularly his dealings with Einstein and Weyl. In addition, Flexner was struggling with tonsillitis and then his tonsillectomy and recovery. Frankfurter was on leave at Oxford during the 1933-1934 academic year, and thus unable to attend the first several Board meetings. While in England, his participation was limited to correspondence with Flexner. When Frankfurter received the October 1933 Board minutes, he reacted to the mention of Weyl's $15,000 salary. Frankfurter inquired as to whether the figure was a fixed scale for all Institute faculty.[17]

Unlike his colleagues, Weyl's compensation package included a stipend of $1,500 to purchase a life insurance policy. At the next trustee meeting Flexner planned to request an additional lump sum payment of $4,500 to defray Weyl's expenses in moving from Germany. Flexner wrote Frankfurter that "the offer made to Weyl is in effect substantially that enjoyed by his colleagues of professional rank. It differs somewhat in form because of his personal and domestic situation and his obligations to his wife's people and his own."[18] The "professional rank" remark was disingenuous in that there were three younger professors whose salaries were bracketed at $10,000. One week later Flexner realized the wisdom of being more forthright with a trustee who was a lawyer. He followed up by listing the salary of each professor and noting distinctions in age and eminence.[19]

As it turned out, Frankfurter was more troubled by another issue raised in the earlier letter. His response went right to the point.

> You would not want me to keep from you the very real anxiety aroused in me by your statement as to the financial treatment of members of the Institute. I fully understand, of course, the existence of categories in your salary scale, a differentiation between $15,000 for older men of high distinction and $10,000 for younger men as plainly rational. But individualization within these categories or outside them seems to me hostile to the underlying assumptions of the Institute and gravely menacing to the realization of its purposes, and this for two reasons.[20]

Variations would be seen as preferential treatment and undermine the scholarly environment. Moreover, such a policy tends to select for the aggressive and greedy.

Frankfurter had witnessed the ill feelings that salary differentials generated among elite scholarly communities. As a clever lawyer, he could anticipate Flexner's defense. While it was not surprising that Frankfurter's letter closed with a pre-emptive strike, the patronizing tone contrasted sharply with the deference Flexner was accustomed to receiving from Beard and the trustees.

> Really, dear Abe, you are in for great trouble and damage to your Institute once you embark upon this path, however much benevolence and justifiable confidence in your own disinterestedness in making differentiations may enter into the process. Really it is idle for you or anyone else to believe that he can make nice discriminations between degrees of eminence in scholarship.[21]

Flexner was confident of his ability to make equitable distinctions in scholarship and need. He set out to win over Frankfurter, adopting an approach that had proven effective with other trustees. Flexner recited his own successful handling of stipends administered by the General Education Board.

Next he alluded to similar salary practices by German universities and by Simon at the Rockefeller Institute. The final paragraph included a platitude about maintaining the Institute as "a paradise of scholars."[22]

The name dropping did not succeed. Frankfurter, conceding his own intolerance, dismissed the German universities as "poisonous centres of anti-Semitism, militarism, and Nazism."[23] The General Education Board experience was irrelevant to "a permanent group of scholars." Frankfurter ridiculed the reference to a "paradise of scholars" as "exuberant rhetoric." He then reiterated his own position. "What I do insist on is that whatever classifications there be—and there ought be very few classes—they should be impersonal." Frankfurter went on to say that Flexner was falling into the trap of an administrator who, among other delusions, sees himself as "Lady Bountiful, or, to keep my sex straight, Kris Kringle." Frankfurter concluded his letter with a fair warning. If Flexner found his candor too problematic for membership on the Board, then now was the time for an annulment.

Flexner was offended by Frankfurter's "bluntness." After seeking reassurance from Aydelotte, Flexner attempted to divert the debate to his strength, German universities. Frankfurter would have none of it: "I am bound to say that your reply to this letter leaves the central point of the communication unattended."[24] It would remain so. Flexner had every intention of customizing compensation to individual circumstances. It was just a fluke that there were currently only two salary levels among the faculty. As the Institute expenditures got closer to its income, more would become necessary. Flexner could neither give in nor hold his own in an argument with Frankfurter. Flexner's letters continued to explore the German university experience.[25]

When Flexner first planned the Institute, he envisaged spirited discussions with trustees. Frankfurter was the first trustee to act out his role as Flexner had originally conceived it. Flexner found reality to be less romantic than abstraction. There remained the question of whether Frankfurter should stay on the Board. Flexner could write off the recent dispute as a random conflict, but it was more likely a precursor to a pattern of unreasonable behavior. Flexner was double bound. Keeping Frankfurter on the Board was likely to impede its harmony. On the other hand, Flexner had touted Frankfurter as a real catch for the Board. It would be embarrassing to explain his resignation to Bamberger, prior to his attending a single meeting. Frankfurter remained on the Board.

The Flexner-Frankfurter quarrel took place during the first few months of 1934. Flexner was 67 years old and feeling his age. The tonsillectomy had destroyed his sense of taste and appetite. For the month preceding the

Frankfurter dispute, Flexner was immobilized by neuritis in his shoulder.[26]
His problems were all compounded with a sleep disorder. Flexner was weary
and looking toward retirement, but his work was not finished. With just one
school, the Institute was a program in mathematics. At least another school
was required to legitimize its mission for advanced study. Flexner believed
that only he could assemble the right men. As Flexner reported at the January
1934 meeting of the trustees, he had one remaining task:

> I realize very deeply that the greatest service that I can perform during my
> tenure of office is that of bringing together the persons on whose abilities
> and devotion the fate of the Institute depends. I have no anxiety now as
> to the School of Mathematics. That is a solid achievement. If I could feel
> that in the next few years, we could do something similar in the field of
> economics and politics, I shall have made the highest contribution that I
> can make to the quality and permanence of the Institute, and I shall be
> easier in mind if two schools are in operation rather than one.[27]

Flexner was counting on one of the new trustees, Frankfurter or Stewart,
to lead the economics school. As the argument with Frankfurter progressed,
Flexner increased his reliance on Stewart for advice. Some of Stewart's ideas
had particular resonance with Flexner. Stewart reasoned that to construct a
radical revision of economic theory required thinkers uncommitted to any
current dogma. This ruled out established economists such as Viner, who
had begun working in the Roosevelt Treasury Department. Viner was eliminated from consideration, even as he continued to receive strong support
from Frankfurter.[28]

Stewart further advised "to discover someone of great ability in the field
of history who would switch to the economic field, which he would enter
with the proper background and perspective and with an absolutely unbiased
mind."[29] While Flexner had not identified any outstanding young American
economists, there was an historian whom he had considered all along for an
appointment to the Institute faculty. Edward Earle was born May 20, 1894,
in New York City. He attended Columbia University, where his education
was interrupted by military service in World War I. After the War Earle returned to Columbia and was appointed a lecturer while pursuing graduate
work in history. Earle received a PhD in 1923. His thesis, published in
book form as *Turkey, the Great Powers, and the Baghdad Railway*, was highly
acclaimed. Earle remained at Columbia and rapidly advanced through the
ranks, becoming a professor in 1926.[30]

In 1919 Earle married Beatrice (Bee) Lowndes. As with the Mitranys, the Flexner and Earle families developed close ties. When Ed contracted tuberculosis, it became impossible for him to teach. Since no tuberculosis drugs were then available, Earle relocated to treatment facilities in the more favorable environments of Colorado Springs and Saranac Lake. During this crisis Abe did all he could to support the Earles. With Bee mostly in New York, Flexner used his contacts to assist her in finding suitable employment. For Ed, Flexner was a faithful correspondent, always offering encouragement.[31]

There was probably a therapeutic element in Flexner's requests to Earle for advice in planning the Institute. However, sympathy would never have been enough for a faculty appointment. Charles Beard confirmed Flexner's high opinion of Earle. When Stewart advocated an historical approach to economics, Flexner decided that Earle was an ideal choice. Flexner raised Earle's name at the April 1934 Board meeting. Earle was then convalescing in Colorado Springs. Trustee Florence Sabin was an expert on tuberculosis who had treated Earle. Flexner proposed to have Sabin evaluate Earle over the summer. If the report were favorable then Earle would receive a one- or two-year appointment at half pay ($5,000) with leave of absence. That the most optimistic prognosis placed Earle's availability one and one half years in the future suited Flexner's timetable. He was expecting to open the School of Economics and Politics in 1935.[32]

Flexner was counting on Stewart to lead the new school. The other faculty would include Mitrany, Earle, and a young economist to be named later. No announcement would be made until Stewart committed. In theory, there remained the flexibility to select a different subject for the second school, but, in his own mind, Flexner was committed to economics. This was apparent in his April message to the Board.

> The School of Mathematics and Mathematical Physics represents an incursion into the field of science. The School of Economics and Politics represents an incursion into the field of the social sciences. That leaves the third great unoccupied territory of infinite cultural importance, namely, the humanities, in which I should include art, archaeology, and music. I am inclined to think that, whenever the means are forthcoming, I should like, before I lay down my directorship, to start with a nucleus in that field. Thus the Institute will have three foci from which it can, as men and means are available, cover the field of intellectual and spiritual endeavor.[33]

The case for a third school was sound. Not only would humanities greatly broaden the Institute's span of scholarship, but an additional program pre-

cluded the unsavory scenario of two schools in direct competition for resources. Three was the minimum number of schools feasible for the Institute.
The difficulty was that the means were not available. The Mathematics budget for the following year was about $130,000. Adding in Flexner's salary and
other administrative costs left barely $80,000 for the two additional schools.
Flexner's entreaties to Bamberger did yield a $1 million donation. It was
earmarked to assist in site acquisition and launching economics. With the
depreciation of Institute assets, however, a portion was required to restore
the endowment to its original magnitude. Flexner's message to the trustees
included a hint to Bamberger of the need for a supplement to support the
program in humanities.[34]

When Flexner alluded to humanities at the April meeting, he gave no indication of the substantial plan that had been developed over the past month.
It was Flexner's custom to withhold names of prospective faculty from the
trustees until the deal was a *fait accompli.* Thus Birkhoff was never mentioned to the Board. The only hint of Flexner's intentions on humanities was
the list of "art, archaeology, and music" omitting literature, philosophy, and
other traditional components.

It was actually art history and archaeology that were at the core of
Flexner's plan. While he had a genuine interest in these subjects, it was opportunity that triggered his move in this direction. The men became available.
Most significant was an outstanding art historian, Erwin Panofsky. Panofsky
was born in Hanover, Germany, in 1892. He proceeded through the German
educational system at a rapid pace, receiving a PhD from the University of
Freiburg in 1914. Panofsky held art history positions at various Germany
universities before settling at the University of Hamburg. He pioneered the
iconological approach to art history.

In 1931 Panofsky received a temporary appointment at New York University. For the next few years he alternated semesters in New York and Hamburg. Then Panofsky was among the Jews losing their positions in Germany.
With a connection already established at NYU, it was natural to seek a more
permanent situation in the United States. The leading American art history
programs were at Harvard and Princeton. Charles Morey was chair of the
Princeton Department of Art and Archaeology. Morey saw Panofsky as an
ideal fit for Princeton. Panofsky was also enthusiastic about the "atmosphere"
for continuing his work at Princeton. A temporary arrangement was made,
but the financially strapped university was in no position to make a long term
commitment. Hoping for the best, Panofsky moved his family to Princeton,
a location providing some accessibility to NYU as well.[35]

Flexner's consideration of Panofsky came about in an indirect manner. It was precipitated by a suggestion to Flexner by Frankfurter from Oxford. Just as their dispute was beginning, Frankfurter proposed an Institute appointment for the Oxford scholar Elias A. Lowe. Lowe was a 54-year-old American who had attained international distinction as a paleographer. Flexner was well acquainted with Lowe from his own earlier residence at Oxford. Late in February 1934 Flexner responded to Frankfurter:

> I know all about his work and his standing, but I do not see how at present the Institute can suddenly launch out in the direction of palaeography, for we have not yet the means nor the material. It may be, however, that Princeton, which has an excellent department of medieval studies, could with the cooperation of either the Rockefeller board or the Carnegie Institution strengthen itself with him. I shall take the matter up with Morey, the head of the department, and if he is favorable, with Dodds the new president, who is a thoroughly nice fellow.[36]

When Flexner did meet with Morey, the discussion went beyond Lowe. Morey pitched both Panofsky and Lowe for the Institute. Flexner, mindful of his own financial constraints, was intrigued. An interview was arranged where Panofsky expressed his desire for an appointment at the Institute. For years Flexner had subscribed to the Gilman philosophy that availability of personnel drove the choice of programs. Panofsky and Lowe offered an opportunity to begin a strong school in humanities.[37]

Unlike mathematics and economics, Flexner had some knowledge of archeology. His undergraduate major was Greek. In 1925 Flexner took a winter vacation to Egypt and Greece. Before it was over he was connecting philanthropists to field projects in these countries, tracking the results with great interest.

The emphasis of the Princeton department was classical and medieval archeology. Flexner and Morey explored possible personnel for the Institute. They found agreement on a slate that, together with the University department, would elevate the city of Princeton to a lofty standing in art history and archeology. It was precisely the sort of partnership with the University that Flexner had desired all along. Moreover, in Morey, he had a willing and capable advisor, like Veblen. It fell to Morey to draft a memorandum providing the rationale and details of how the Institute might proceed. He identified the

lacunae which have made themselves insistently felt within our local research
. . .

(1) a specialist in the later Middle Ages and the early Renaissance, a "Quattrocentrist" in short, whose preference for Italian or Northern Renaissance would be immaterial, provided he would bring to bear upon our work an outstanding competence in the period, and the critical acumen and ability to synthesize the diverse phenomena of the end of the Middle Ages. (2) For our work in the illustration and illumination of manuscripts, we need the constant help of a palaeographer of demonstrated authority. (3) With the exception of Elderkin, our classical archaeologists are field-workers and likely so to continue. The research personnel at home could be profitably enlarged by (a) a specialist in Greek architecture (with special reference to the excavation of the Athenian Agora), and (b) by another scholar of outstanding competence in Greek epigraphy. It is hardly necessary to consider numismatics, since Princeton has already a competent numismatist in Weber, and a ready consultant in New York in the person of Newell, who is probably the best scholar in this field. (4) A Near Eastern archaeologist with a special competence in Islamic art. The recent development of Islamic literature and history within the Department of Oriental Languages and Literatures would provide such a scholar with the necessary assistance in the philological and historical aspects of his subject; his presence in the archaeological group would be immensely helpful, not only for the interpretation of the finds at Antioch, but for the constantly recurring problems of Islamic influence which often baffle the students of the mediaeval art of Europe.[38]

To fill the lacunae Morey went on with the

Desiderati . . . The most brilliant scholar in the late Middle Ages and early Renaissance that we know is Erwin Panofsky, until recently Dean of the Philosophical Faculty of the University of Hamburg, and at present Visiting Professor of Fine Arts in New York University, who will be making his home in Princeton during the next year and a half, at least. He meets in every way desideratum no. 1, and as he seems to be available, this particular problem can be solved in a highly satisfactory way.

For our palaeographer, one would like to find a scholar combining high proficiency in both Greek and Latin palaeography, but such a genius does not, to our knowledge, exist. The most outstanding scholar in Greek manuscripts that we know is Mgr. Giovanni Mercati, prefect of the Vatican Library, who is obviously not available. In this country we have Kirsopp Lake at Harvard, an excellent scholar, but not in the same class with Mercati; as to his availability I would not be able to speak. For Latin palaeography one thinks at once of E. A. Lowe, now at Oxford, the best of our American scholars in the field, and second to none in Europe; his name would immediately associate itself in Princeton minds with desideratum 2.

In the case of desideratum 3a in Classical Archaeology, the obvious person is Dinsmoor of Columbia, a scholar possessed of intuition amounting to genius in his preferred field of Greek architecture, and peculiarly fitted to fill out our group in this subject. If he came to the Institute, it would greatly facilitate the completion of his monumental work on the Propylaea of the Acropolis, and he would have in Stillwell of our staff a younger coadjutor who promises to measure up to Dinsmoor's competence as he grows and expands. For 3b, Meritt of Johns Hopkins, known for his brilliant work on the Agora inscriptions, is an outstanding epigraphist, and hardly to be bettered by any candidate in Europe.

As to an Islamic specialist, we would have no recommendation to make at present, since the search would have to be made, in view of the age of the outstanding scholars in this field, in the ranks of the younger generation. This need is in fact not so pressing as the foregoing, and could await an extended sifting of personalities in this country and particularly in Europe.

Although Morey's memorandum portrayed the Institute faculty in a supporting role (filling university lacunae), the slate of Panofsky, Lowe, William Dinsmoor, and Benjamin Meritt stood on its own as a stellar school. Each of the nominees was an outstanding scholar. Since the Institute was not offering a degree program, it had the luxury of selective specialization. Moreover, both Panofsky and Lowe were well disposed toward joining the Institute.

The nomination of Meritt may have originated with Flexner. Flexner knew him well. Meritt's in-laws were neighbors in their summer homes on Lake Ahmic in Canada. Ben Meritt was born in 1899 in Durham, North Carolina. He received a PhD in classics from Princeton in 1924. Meritt quickly established himself as one of the most promising people in the classical field. His specialty was epigraphy, the study of ancient inscriptions.

In 1932 Meritt was offered a professorship at Johns Hopkins. His current position at the University of Michigan entailed a number of commitments for the following year. Meritt sought advice from Flexner, who intervened on his behalf. Flexner urged the Hopkins president to defer the appointment for a year. Satisfactory arrangements were made and Meritt was on the Hopkins faculty when his name arose in the discussions between Flexner and Morey.[39]

The memorandum from Morey was completed two weeks prior to the April Institute Board meeting. In one month Flexner had made more progress on humanities than he had in over three years on economics. Even so, Flexner would not consider moving humanities up in the queue. The message to the trustees was that economics was next. Humanities would have to wait until additional money was available.

Benjamin Meritt and Abraham Flexner in Canada in the 1920s. (Courtesy of the Archives of the Institute for Advanced Study.)

While economics was to be Flexner's priority in the summer of 1934, he moved forward cautiously on the other schools. One of the difficulties in a tight budget was that archeology required staff, infrastructure, and space. Included in Morey's memorandum were specifications for an addition to the Princeton archeology building. Flexner sent a noncommittal query to Lowe, seeking to determine his needs for carrying on his work in Princeton. For the other humanities prospect, Panofsky, Flexner set out to prepare some delicate ground. He met with the chancellor of New York University to ascertain the university's intentions toward Panofsky. NYU had hopes of firming up its current limited arrangement. Knowing that Panofsky preferred the Princeton location, Flexner came away optimistic about the outcome, but fully mindful that "we must be punctilious in helping him to carry out his arrangements with New York University."[40] It was at this time that Morse entered the picture for a possible appointment in mathematics.[41]

The entire economics strategy still depended on Stewart. It appears that Flexner intended to shadow Stewart, waiting for a timely opportunity to win him over. During May, Flexner kept his summer schedule open. Under consideration was a trip to England for consultations on economics. In June,

Flexner suddenly decided to make the voyage. Rather than crossing with his wife and daughter, Flexner departed a week later on the same ship as Stewart.[42]

On the voyage they discussed personnel and plans for the economics school. Stewart declined to accept a faculty appointment, but did not foreclose the possibility in the future. Flexner made it clear that a professorship would always be available for him. When they reached England, Flexner relied on referrals from Stewart to augment his search for economists. After a few weeks of meetings in Oxford and London, Flexner sent an upbeat progress report to Bamberger:

> I am beginning to feel the way I felt in the early days about mathematics, namely, that I know with whom to start and one or two persons besides, but I am keeping my mind open, for it is always possible up to the last moment that somebody else may quite unexpectedly turn up.[43]

Despite the brave face, Flexner did not have an economist.[44]

It had been an exhausting year and the trip to England deprived Flexner of his annual retreat in the Canadian woods. Hoping for some relief from the after effects of his tonsillectomy, Flexner took a long vacation to relax and indulge in the curative baths at Gastein and Aix-les-Bains. Well rested, he departed from Europe in September.[45]

Throughout the summer Flexner worried whether any initiatives were advisable in view of the hopeless American economy. Unemployment remained astronomically high. The New Deal and its hyped initiative, the National Recovery Administration, thus far had failed to abate the depression. Some members of Roosevelt's team had already left the administration.[46] Just before returning home Flexner summarized his concerns to the paleographer Lowe:

> Unless Roosevelt still further muddles our financial situation, it ought to be possible to carry out the plans which we have tentatively discussed, but at the moment I am completely mystified partly because I cannot understand the report which comes from America and partly because European financiers and publicists, with whom I have talked, are utterly bewildered . . . . I cannot conceive at what Roosevelt is driving either in his monetary policies or his labor policies, and it would be wrong for me to take any step beyond those already taken until the atmosphere has cleared. We seem to have reverted back to the Bryan-Cleveland era, only there is no Cleveland. At the bottom of my heart I believe that Roosevelt is in intelligence still a Harvard sophomore and the country will find it out long before his term expires.[47]

Facing Flexner was an overflowing plate of possible moves to propose at the upcoming Board meeting. He had met with Lowe in England and confirmed his availability. Flexner was now referring to the third school as the School of Humanistic Studies. With Lowe and Panofsky a small, but auspicious, beginning was possible in the following year. Meritt and others could be added later. Flexner was anxious for the Institute to reach the humanities and certain of the soundness of the plan.[48]

The School of Mathematics was already a huge success. Marston Morse had expressed interest in joining the faculty. With Birkhoff and Lefschetz out of the picture, Morse was the only remaining American mathematician with the stature to enhance the group. Flexner wanted to create a sixth position for Morse.

At the top of Flexner's to do list was the School of Economics and Politics. The only name that was ready to go forward was Earle, who had received conditional approval at the previous meeting. Since Flexner was hoping to propose an economist or two at the January meeting, it was essential to reserve sufficient budgetary resources to cover this eventuality.

The long-deferred issue of site acquisition was another competitor for funds. With new schools on the horizon, infrastructure was a pressing consideration. The School of Economics and Politics needed a home. University facilities could accommodate only a small number of humanistic studies faculty, and Fine Hall had reached capacity with the large influx of Institute mathematicians. Constructing a campus required long-range planning. If the Institute intended to occupy its own buildings in the next few years, then it was incumbent to purchase the land.

Whenever Flexner faced prioritization, men came before buildings. Even so, there were insufficient funds to move on economics, humanistic studies, and Morse. Flexner's commitment to economics was his highest priority. Of the remaining two considerations, Morse was ahead of humanistic studies. At the October Board meeting Flexner would seek approval of Earle and Morse.

Flexner was disregarding his own fundamental principle. Lowe and Panofsky were both first-rate and available. The case for humanistic studies was especially compelling when contrasted with the dim prospects in economics. Flexner was yet to enlist a single economist to work on the grand revision of economic theory. Contributions by the bed-ridden historian Earle were severely limited for the foreseeable future. This left the political scientist Mitrany, with whom Flexner's correspondence indicated a growing estrangement. The School of Economics and Politics was crying out for a redesign.

Despite his boasts of the Institute's plasticity, Flexner would not back off from economics. He was determined to impress Bamberger with the relevance of the Institute.[49]

Flexner's recommendations had never been challenged by the trustees. The October 1934 meeting had some potential for controversy. It marked the first attendance by Frankfurter and another new trustee. Herbert Lehman had resigned from the Board following his election as governor of New York. Veblen was chosen as Lehman's successor, after Flexner failed in his nomination of Eisenhart. Both Veblen and Frankfurter were opinionated and persuasive advocates. Veblen's mathematical partisanship was a concern for Flexner, who did not know what to expect from Frankfurter. The two new members were unlikely to take issue with Flexner's desire to move on economics and Morse over humanities. Frankfurter was acting as a proponent of the social sciences and Veblen did have a bias toward mathematics. There was nobody to stand up for art history.[50]

Personnel was an area of likely sensitivity to Veblen and Frankfurter. Here Flexner maintained his caution in floating names before the Board. Only Earle would be mentioned at the October meeting. Neither Panofsky nor Lowe was named. To a direct question from Frankfurter, Flexner replied that he was unprepared to submit any economist for consideration by the Board. Flexner did ask for authorization to proceed on a sixth mathematician, "preferably an American," withholding identification of Morse as the intended candidate.[51]

No trustee possessed any basis to compare the opportunities for personnel in the three schools. What should have been apparent was the disproportionate funding for mathematics, and its likely limitation on other initiatives. Maass was well aware of this contingency, but he preferred to express his concerns privately. As treasurer, the accountant Leidesdorf must have considered the ramifications.[52]

At the October meeting, Flexner acted pre-emptively to allay concerns over the mathematics budget. He pointed out that the School's new professorship should be viewed as a redirection of the money already allocated temporarily for Dirac. While future visiting professorships were certainly desirable, an additional mathematics position could be achieved, if necessary, without significantly encumbering additional funds. With this ploy Flexner succeeded in diverting attention from budget priorities. However, the meeting did include a departure from Flexner's customary tight control of the script. A discussion was opened on the direction for the program in economics. It was a rare call for trustees to express a substantive opinion. Frankfurter and Stewart were in agreement on the advisability of an historical attack by a small group of young, objective scholars with a common purpose. The discussion placed no restrictions on Flexner.[53]

Following the meeting, Flexner moved forward on Morse. In their spring discussion Flexner had informed Morse that "as soon as this financial situation cleared up I desired to place all professorial salaries at $15,000."[54] Meanwhile Flexner assured the Board that the new salary would approximate the $10,000 being paid to Dirac. With little perceptible change in the financial situation, Flexner decided to split the difference on Morse. The $12,500 salary exceeded Harvard's counteroffer.[55]

When Morse agreed to the terms, Flexner polled the executive committee for final approval. He reported that Morse was the choice of the mathematics faculty. Also included in the canvass was a seemingly routine request to bring in economist Jacob Marschak for a month-long visit. Marschak was a European Jewish refugee in his mid-thirties. Forced out of the University of Heidelberg one year earlier, he became a Fellow at Oxford. Marschak's strong mathematical orientation was unusual for economists of his time.[56]

Marschak came to Flexner's attention earlier in the year in an unsolicited nomination from Schumpeter. The astute Schumpeter praised Marschak's quantitative acumen and his recent work on elasticity of demand. Flexner interviewed Marschak on his summer visit to Oxford. The impression was favorable, but inconclusive. When Flexner returned to Princeton, he asked von Neumann's advice on the economics candidates who employed mathematical techniques. Von Neumann placed Marschak above all the others, suggesting that a personal interview would permit a more definitive judgment.[57]

With Frankfurter among the nine Executive Committee members, unanimity on the two issues could not be taken for granted. Surprisingly, Frankfurter overlooked the creation of a new salary bracket for Morse. He enthusiastically supported the appointment, noting his satisfaction that the mathematics faculty were included in the process. It was the invitation to Marschak that drew Frankfurter's protest. He was shocked that an economics name suddenly had arisen in the two weeks since the Board meeting.[58]

Frankfurter objected to the Marschak invitation on two grounds. Certain aspects of Marschak's candidacy required more discussion, and the nature of an extended transoceanic visit carried serious implications. Frankfurter suggested that the proposed action were "hurried" and urged that it be put over until the next Board meeting.

Flexner insisted that there were no implications. Several economists were being invited for individual visits. In particular, he was about to lunch with a Federal Reserve Board economist whom Stewart had proposed for consideration. Flexner assured Frankfurter that, in both cases, the arrangements were being made in a manner that entailed no future commitments. Embedded in Flexner's reply was a careless sentence: "I don't know that, dealing as we are with young persons, it makes much difference with whom we start."[59]

Frankfurter elaborated his concerns. Marschak (whose wife was also to be invited) was certain to attach expectations beyond what might accompany a luncheon invitation. If no position were forthcoming, there could be damage to Marschak's sensibilities. Moreover, Oxford was doing its best to accommodate refugee scholars. With Marschak's visit to the Institute, Oxford might well give preference to others who did not have prospects abroad. Under these circumstances the Board would feel an obligation to offer an appointment to Marschak.[60]

As for Marschak's suitability, Frankfurter was skeptical of the need for a mathematical economist on the Institute team. He was also troubled by the prospect of appointing a foreigner. Frankfurter urged that such matters should receive thorough Board discussion prior to the issuance of any invitation. Another Flexner-Frankfurter conflict was underway. With each letter the acrimony escalated. Frankfurter took every opportunity to insist that the choice of economist did make a difference.

On the uncertainty of how Oxford might react to Marschak's Institute invitation, Flexner deferred to nobody in his understanding of European protocol. Frankfurter was striking some sensitive areas. Flexner himself had a policy of strongly favoring Americans over foreigners. He agonized over maintaining a proper foreign-domestic balance. In view of the Institute's nondiscriminatory ideals, Flexner had deliberately concealed these considerations from the Board. There was no chance of Marschak's nationality reaching an agenda.

Flexner and Frankfurter couched their arguments in terms of the role and responsibilities of trustees. It was all pretense. Neither man had much respect for the Institute Board's capacity to evaluate an economist. Frankfurter believed in the principle of faculty selecting their colleagues. Since there were no economists on the faculty, the task devolved to qualified trustees, meaning himself and Stewart.

To Flexner, devising the modalities for faculty selection were among his prerogatives as Director. Flexner's approach was to rely on his own instincts and the advice of a dominant consultant such as Veblen or Morey. When

Frankfurter agreed to serve on the Board, he was functioning in this role for the School of Economics and Politics. The arrangement ended with the argument over salaries earlier in the year. Stewart was now Flexner's economics advisor and Frankfurter was out of the loop. There is no record of when and how Stewart reacted to the Marschak invitation. It is inconceivable that he was not consulted prior to the Executive Committee poll. Since Stewart believed that economics was in need of a more scientific approach, a mathematical economist suited his outlook.[61]

At the heart of the Frankfurter-Flexner dispute was a delicate matter. It was their respective perceptions of Flexner's qualifications to make judgments of scholarly potential. Flexner knew that his own writings and honorary degrees, like those of Gilman, did not measure up to the research of the people he was seeking. Flexner was still comfortable as an arbiter of intellectual greatness, but insecurity persisted just below the surface. Most of the trustees were unaware of the chasm between Flexner's scholarship and that of faculty such as Weyl. Among the more academically-oriented Board members the subject was largely taboo. Aydelotte was a consummate diplomat, and Veblen aggressively pushed his agenda without explicitly challenging the Director's bona fides. Frankfurter did not observe these niceties.

Flexner made the mistake of using a poor analogy to support his authority in making faculty appointments. A letter to Frankfurter reasoned,

> surely, no Board of Trustees will in the end interfere with my responsibility when it comes to a final choice. How would the Harvard Law faculty feel if, when a recommendation has been made by the faculty, the corporation substituted somebody else?[62]

Frankfurter asked incredulously whether Flexner believed that "the analogue of the relation of the Harvard Law School faculty to the Harvard Corporation is your relation to our Board."[63] The attack hits its mark. Flexner was insulted and furious. He belatedly moved to close the argument by proposing that they agree to disagree. Frankfurter responded with disdain.[64]

So much rancor was triggered by the simple question of whether to bring in Marschak for a visit that his candidacy took on secondary significance. An invitation was never extended. As the dispute was heating up, Flexner backed down, citing deference to the minority view of Frankfurter. Of greater bearing was the emergence of an American economist. Stewart set up a luncheon meeting for Flexner with Winfield Riefler. Flexner had had prior discussions with Riefler, identifying him as an impressive young economist with a desirable balance of theoretical knowledge and practical experience. Moreover, he was an American. Flexner broached the possibility to Riefler of an Institute appointment.[65]

Riefler was born in Buffalo, New York, in 1897. His undergraduate work was at Amherst while Stewart served on the faculty. Riefler received a PhD in 1927 from the Brookings Graduate School (that later evolved into the Brookings Institute). For ten years Riefler served on the staff of the Federal Reserve Board's Division of Research and Statistics. He was currently chair of the Central Statistics Board and an economic advisor to both the President's Executive Council and the National Emergency Board. In 1930 Riefler published a book entitled *Money Rates and Money Markets in the United States*.

One week after their luncheon, Flexner requested another meeting with Riefler. The purpose was to ascertain whether Riefler were disposed toward joining the Institute faculty. When Riefler responded favorably, Flexner asked him to prepare a project description. Riefler drafted a ten-page "Proposed Economic Unit of the Institute for Advanced Study." The memorandum set forth Riefler's vision for the economics school to address the depression problem that Flexner had originally set for it.[66]

Riefler contended that economic theory was predicated on forces that existed prior to significant industrial development. Perishable goods such as food and clothing were then the dominant segments of production and consumption. With the industrial boom automobiles, machines, and other durable goods assumed a prominent role in the economy. The nature of durable goods made their demand substantially more flexible than that of perishables. The impact on markets was both considerable and poorly understood.

Riefler proposed that the School of Economics mount a full-scale study of durable goods. The investigation would be, by Institute standards, personnel intensive. Riefler's implementation called for "a group of economists of outstanding ability" and "a small statistical research and clerical staff." Flexner was undeterred by the expense of the statistical methodology. He immediately sought an evaluation from Stewart.[67]

Stewart's response survives among the Institute records, providing a rare glimpse into the thinking of Flexner's enigmatic advisor. Unlike the now *persona non grata* Frankfurter, Stewart could deliver a negative recommendation without deprecating the candidate.

> I am very much impressed with the memorandum Riefler sent you. It seems to me a cogent and effective presentation of his case and I am persuaded that in making a start, it is probably wise to select some field of interest and use

the problems of that field as a basis for selecting personnel and of establishing some unity in the work. Whether the problem which Riefler outlined is *the* problem is another question. From the form of his memorandum, I judge that with him it is a matter of "love me, love my problem". In this decision he may be wise though it forces us to a decision as to whether we want both him and his problem. In economics, my preference runs toward someone who is possessed with *some* concrete problem but who is prepared to deal with its general implications. This seems to me to furnish the best hope of escaping from the vagueness of superficiality which has affected so much current work in economics and of establishing a fresh approach.[68]

Stewart's disapproval did not deflect Flexner in his pursuit of Riefler. No doubt Flexner was feeling pressure to appoint an American economist, but it is difficult to reconcile his captivation with Riefler, particularly in view of the previous dismissal of Viner. Riefler's bureaucratic and statistical background hardly portended a bold reconstruction of economic theory. In placing Riefler ahead of Viner and Marschak, as well as Panofsky and Lowe, Flexner was giving extraordinary weight to his own instincts and impressions. Nevertheless, he was set on Riefler as the next member of the Institute faculty.

There was another difficulty with Riefler. His project was unsuited to the interdisciplinary social science approach for which Mitrany and Earle were recruited. Flexner had only recently informed Mitrany and Earle of each other's appointment. Now he sent them copies of the Riefler memorandum, attributing it to an unnamed, young economist. Mitrany and Earle reacted similarly. Both regarded durable goods as an important problem, but they were deeply troubled by the prospect of their entire school becoming committed to Riefler's approach. While Earle and Mitrany were delighted at the prospect of becoming each other's colleague, neither had any interest in pursuing durable goods or employing a statistical methodology.[69]

Earle was in the midst of an unbearably painful treatment of his tuberculosis. Bedridden, he urged Flexner that the Institute not "stake all its chips on this one card."[70] Flexner responded with compassion for Earle's ordeal and assurance that both he and Riefler would receive free rein. Smoothing the Riefler memo over with Mitrany was less of a priority. Relations between Flexner and Mitrany were already strained.[71]

Over the past year Mitrany had sought Flexner's support for several initiatives. It began with Mitrany's request to retain the services of the graduate student who served as his research assistant at Harvard. When Flexner declined, Mitrany pleaded for reconsideration. Mitrany's recommendation for a faculty appointment in psychology met a similar fate. Mitrany was, in part, a victim of his Harvard and Oxford associations with Frankfurter. A few common threads are detectable in the letters from Mitrany and Frankfurter to Flexner. Among these was a proposal to bring in several faculty candidates and observe them together over an extended period. Flexner explained that he could not sanction competitive auditions for faculty positions. Flexner's paranoia emerged during the argument with Frankfurter over Marschak. In a parallel correspondence Mitrany alluded to the Director's intention to hire a mathematical economist. Fearing that Mitrany and Frankfurter were colluding behind his back, Flexner challenged Mitrany to furnish the source for his mathematical economist allegation.[72]

While Mitrany was declining in Flexner's estimation, Frankfurter had appeared to reach bottom. Flexner regarded Frankfurter as a "trouble-maker," declaring to another trustee, "I have no intention of ever conferring with him or dealing with him . . . in any manner whatsoever."[73] With a Board meeting in three weeks, some sort of showdown was inevitable. Flexner intended to marginalize a person who could not be intimidated. The question remained as to when Frankfurter would strike.

Flexner's main objective for the January 1935 meeting was to obtain trustee approval for the appointment of Riefler. After several weeks of courtship, Riefler had agreed to a salary of $12,000. The $2,000 differential over Mitrany and Earle was certain to arouse Frankfurter's objection, but where would he stand on Riefler as the first Institute economist? On the surface, Riefler appeared to be Stewart's nominee. In fact, Stewart had serious reservations, but neither he nor Flexner had any intention of disclosing these doubts to the other trustees. Under this misapprehension, Frankfurter supported the appointment, but balked at the salary discrepancy. Frankfurter was adamant that all three School of Economics professors receive the same compensation, whether it was $12,000 or $10,000.[74]

In a time of severe deprivation, $10,000 remained a lofty salary for a university professor. Frankfurter's egalitarian ideals were out of step with the beliefs of Bamberger and the other businessmen on the Board. The only support came from Veblen, who, for some time, had urged Flexner to equalize the mathematicians' salaries at $15,000. Veblen suggested that the Board endorse a policy of equal salaries as a future objective. Seeing no bandwagon effect, Veblen elected to retreat.[75]

Frankfurter soldiered on alone, insisting on a recorded vote. The outcome established a precedent. Frankfurter became the first trustee to have a dissent noted in the minutes. Other business went more smoothly. The trustees accepted Flexner's proposals to defer humanities and to authorize, for a second year, the visiting professorship in the School of Mathematics. While no name was provided, Wolfgang Pauli would replace Dirac for 1935-1936, joining the six permanent professors.

For Flexner, the meeting was a success. Not only was his entire agenda adopted, but the charming and erudite Frankfurter had revealed his belligerent side. When the next dispute flared up, as it inevitably would, Flexner was well positioned. The founders would recognize the threat posed by Frankfurter to Board comity. The confrontation came sooner than expected. Frankfurter persisted on the salary issue. As Flexner telegrammed Riefler, "Board of Trustees ratified with great enthusiasm your appointment yesterday on basis upon which we agreed,"[76] Frankfurter composed his own letter:

My dear Riefler:

Ever since I have been on this faculty, for now a little over twenty years, it has been my practice to tell acquaintances whose names have come up for our consideration directly what doubts or difficulties I may have had to raise in faculty meeting. This avoids misunderstanding through the dangers of misreport, however innocent through indirect transmission. That practice of candor seems to me equally appropriate for you and me in the case of the Institute for Advanced Study.

Therefore, I should like you to know that I welcomed your accession to the Institute and voted for it with pleasure and hope. But I voted against the stipend proposed by Dr. Flexner, not because it was too high, but because it was higher than that given to your colleagues in the School of Politics and Economics. For I deem inequality of treatment among men of substantially similar age and scholarly distinction as inimical to the aims of a society of scholars. This is not the occasion to argue the matter, I simply wanted you to know precisely what my attitude was towards your coming to the Institute and to the conditions of your coming.

If you have to leave government—I cannot conceal my regret that you are doing so, in view of my interest in a permanent civil service—I am at least happy that you are giving yourself to scholarship.[77]

Riefler was already conflicted over leaving public service. It was difficult for him to reconcile Frankfurter's unorthodox disclosure with the "great enthusiasm" Flexner attributed to the trustees. Understandably disturbed and puzzled, Riefler proposed that the Board reconsider his case. When Flexner received his copy of Frankfurter's letter, the Director became apoplectic. Frankfurter was jeopardizing a vital appointment. Moreover, he had violated what Flexner believed to be the confidentiality of trustee proceedings.

The gloves came off. Flexner wrote Frankfurter that "your letter to Riefler was a piece of unmitigated impertinence and makes it absolutely impossible for you and me to collaborate in any enterprise whatsoever."[78]

Both Flexner and Frankfurter were approaching the end of their multi-year terms on the Board. While reappointments were taken for granted, formal action was required by a Committee on Nominations. In his letter to Frankfurter, Flexner made his own disclosure. He would inform the Committee of his refusal to be nominated with Frankfurter. "They shall have to choose between us."[79]

Flexner's immediate priority was to mollify Riefler. When this was accomplished Flexner saw an appealing opportunity. He would replace Frankfurter with Riefler on the Board. Together with Veblen there would be a nontrivial faculty representation. Moreover, Riefler immediately would see himself as an integral part of the Institute. Flexner was linking the faculty-trustee positions for Riefler in the reverse order that he devised for Frankfurter.[80]

Meanwhile, an unrepentant Frankfurter attempted to rally support from other trustees. Aydelotte was chair of the Committee on Nominations. In their prior disputes both Flexner and Frankfurter had appealed to Aydelotte for reassurance. In those instances Aydelotte had attempted to avoid condemning either party.[81] Neutrality was no longer possible and Aydelotte made his position clear to Frankfurter. The letter to Riefler was a serious breach. "Trustees, however they may differ among themselves, should communicate with members of the Faculty on questions of this kind through the Director and not over his head."[82]

Frankfurter did receive encouragement from trustee Lewis Weed of the Johns Hopkins Medical School. Weed actually opposed the principle of standardized salaries, but he was fed up with Flexner's domineering tactics. Weed urged Frankfurter to continue challenging the "Flexnerian positions."[83] Nevertheless, Flexner held the high ground, both strategically and on the issue of communication with Riefler. Aydelotte was dispatched to meet with Frankfurter and urge him to resign or to withdraw his name from consideration. Frankfurter considered the request and declined, explaining to Aydelotte: "A resignation now would imply a confession of wrong-doing which I do not in the slightest feel."[84] Flexner, and not Frankfurter, was selected for renomination.[85]

A formal announcement of the School of Economics and Politics followed the Board's confirmation of Riefler in January 1935. It had been nearly two years since Mitrany's appointment. During this period the school existed in a stealth mode as Flexner sought its guiding professor. With Riefler on board, Economics was established and Flexner was ready to commit publicly. The next step was clear. A third school in the humanities was necessary to complete the foundation for the Institute for Advanced Study.

Flexner liked the name Humanistic Studies. It provided a suitable home for his first intended professors, Panofsky and Lowe, and yet left room for future expansion into philosophy, literature, and other subjects. Whether there were funds to accommodate Panofsky and Lowe was the immediate issue. Both had been waiting for a year while Flexner cautiously delayed their consideration so as not to jeopardize the start up of the School of Economics and Politics.[86]

An urgency to proceed on Humanistic Studies came about early in 1935 when both Panofsky and Ben Meritt received substantial offers from other institutions. Meritt sought advice from Flexner on whether to leave Johns Hopkins for the $9,000 salary and release times perks available at the University of Chicago. Flexner was not a disinterested party. He had his own designs on Meritt. Flexner had planned on Meritt gaining stature for a few more years at Hopkins and then joining a second wave of appointments in the School of Humanistic Studies. The handsome Chicago terms, to an archeologist still in his mid-thirties, caught Flexner by surprise. Especially troublesome was the thought of Meritt landing at Chicago, whose president, Robert Hutchins, figured prominently on Flexner's list of misguided administrators.[87]

Flexner reacted quickly to the Chicago offer. He arranged a meeting with President Dodds to strategize over bringing Meritt to Princeton. It soon became clear that no position for Meritt was to be forthcoming from the university. This left Flexner with the dilemma of either losing Meritt or of advancing the opening date for the School of Humanistic Studies. Flexner went to New York and conferred with the Institute Board's Treasurer, Leidesdorf, about the budget. Income for the coming year was projected at $300,000, comfortably covering the $225,000 in anticipated expenditures. It was easy to argue that the margin was more than sufficient to accommodate the salaries of Meritt, Panofsky, and Lowe. There were, however, a number of other considerations.

Only the School of Mathematics had reached a mature state. Its budget included $40,000 to support members, assistants, and outside speakers. No such lines were currently provided for the other two schools. Then there were future faculty. More professors would be needed in both Humanistic Studies and Economics, where Stewart was holding a standing offer for a high-salary position. Another item was the statistical staff that had been promised to Riefler. Just a little arithmetic was required to foresee approaching shortfalls.

Flexner, who had been so critical of universities' poor planning and deficit spending, went into denial mode. He was certain that more money would be forthcoming. If necessary, economies were possible in the School of Mathematics. The visiting professorship could be eliminated and grants for members could be scaled back. His reasoning was disingenuous. Even if these contingencies were adequate, which they weren't, none of this accounting provided for the biggest budget item on the horizon. A committee of trustees was actively considering tracts of land for an Institute campus. This purchase and future construction were likely to reduce the endowment. When confronted with such possibilities Flexner invariably lapsed into refrains about Gilman and "men above buildings."[88]

Flexner did make one concession to the budget exigencies. He decided to defer the appointment of Lowe. At the April Board meeting Flexner planned to seek approval to begin Humanistic Studies with Meritt and Panofsky. The choices of Meritt and Panofsky over Lowe were motivated by market considerations. Panofsky also had other opportunities.[89]

For several years, Panofsky had split his time between his home base of Hamburg and a temporary position at New York University. With Hamburg out of the picture, NYU was seeking to make their arrangement permanent and full-time. Flexner knew that Panofsky preferred a position in Princeton. The negotiations were particularly sensitive. Institute trustee Percy Straus believed that NYU, with whom Straus also had connections, was entitled to Panofsky's services. Flexner delayed any formal move until after Panofsky declined an offer from NYU. By then, other institutions were entering the bidding.[90]

Panofsky and the School of Humanistic Studies were the topics of extended discussion between Flexner and Straus. Flexner emphasized the chronology that he had moved to secure Panofsky for the metropolitan New York area only after Panofsky's rejection of the NYU offer. Flexner's carefully worded explanations omitted his collusion with Morey. Panofsky was aware that Flexner would not act while an NYU offer was on the table.[91]

Panofsky was just a part of Straus' more general concerns. Straus made sound arguments that the current budget was insufficient to sustain three mature schools. He pointed out that development of the School of Economics was certain to require substantially more funds, a capital outlay for buildings was approaching, and no significant infusion of funds would occur until after Bamberger's death. On the last point Straus may have been more knowledgeable than Flexner. He was certainly more realistic. Flexner conceded little in debate, but his actions reveal a new attitude toward the budget. Flexner suddenly abandoned his objective of elevating faculty salaries. The compensation proposed for Meritt and Panofsky was $9,000 and $10,000, respectively, merely matching their opportunities elsewhere.[92]

Both Straus and Flexner made their cases to the trustees in April. In the Report of the Director Flexner rehearsed the argument for starting the School of Humanistic Studies:

> (1) I am convinced that the most important contribution which I can possibly myself make to the development of the Institute is in the finding of men, for I have been fortunate beyond most persons in an experience extending over more than a quarter of a century which has brought me into contact with scholars in almost every field of academic endeavor; (2) we must be guided altogether by the possibilities of obtaining men of first-rate importance.[93]

The two right men for humanistic studies were now available and the window would soon close. Straus suggested that it was unwise to begin Humanistic Studies until the long-term needs of Mathematics and Economics became clear and were provided for. The substance and manner of Straus' arguments were more effective than those of Frankfurter at the previous meeting. In the end, the Flexnerian position carried all of the votes, excepting Straus who graciously abstained. Still, Straus had made some points and the debate was a bit too close for Flexner's comfort. On the next day Flexner went to the Rockefeller Foundation, seeking outside funds for the Institute. He came away optimistic, but without any commitments.[94]

With the establishment of the second and third schools, Flexner had realized his dream. Even if there were long-term financial risks, the Institute was a new paradigm for promoting basic research at the highest level. For the first time in America, an unprecedented environment existed for scholars in each of the fundamental areas of science, social science, and the humanities. The Institute would exert a significant impact on the intellectual scene. To be sure, Flexner had not accomplished this by himself. Just as Gilman was aided by Johns Hopkins, Sylvester, and Welch, Flexner was helped by Bamberger, Fuld, Veblen, Morey, and even Hitler. Nevertheless, the Institute for Advanced Study was primarily the result of Flexner's vision and efforts. For this accomplishment Flexner deserves a place with his hero Gilman in intellectual history.

When comparing himself to Gilman, Flexner liked to point to their insight in evaluating men. Gilman's triumphs included the young physics instructor Rowland, the unemployed Sylvester, and Welch. While the Institute obtained a galaxy of stars, it could hardly be attributed to Flexner's assessment of talent. Einstein's genius was no secret, and it was Veblen who identified von Neumann, Gödel, and most of the other mathematicians. For the School of Humanistic Studies the selections were shaped as much by Morey as by Flexner.

It was in the School of Economics and Politics where the staff was most reflective of Flexner's judgment. He cultivated Mitrany and Earle from the start. Viner was the choice of the economics experts, but Flexner decided on Riefler. How did this slate compare to those of Veblen and Morey? The School of Economics and Politics was already in serious trouble prior to its opening. Earle's health remained in doubt and Flexner himself believed that Mitrany was a mistake. In May 1935 Riefler asked about being released from his commitment. The Social Security Act was a few months from becoming law. Riefler was offered a high salary and large budget to play a leading role in implementing the new program.[95]

Flexner's response to Riefler was markedly different than it had been to similar requests from Birkhoff and Weyl. Flexner belittled the Social Security position as a dull administrative post unworthy of serious intellectual engagement. Only grudgingly he conceded that Riefler could be released to pursue this mistaken course: "I do not want to let you out either with honor or anything else. On the other hand, neither now nor at any future time would I stand in the way of any man who really saw a bigger and more congenial opportunity."[96] Flexner's determination to retain Riefler was indicative of his concerns about the school and his conceit that Riefler was a brilliant choice. Riefler elected to stay with the Institute, but his skills were more suited to government service.[97]

Charles Rufus Morey, 1924–1945 chair of the Department of Art and Archaeology at Princeton University. (Courtesy of Visual Resources Collection, Department of Art and Archaeology, Princeton University.)

CHAPTER EIGHT

COMPETING FOR RESOURCES

On October 1, 1935, the Institute for Advanced Study began its third year of formal operation. Recruitment over the past year had nearly doubled the size of the faculty. The School of Mathematics was complete, and the other two schools opened in what Flexner characterized as an "embryonic" state.[1] For the remainder of his directorship Flexner would tenaciously pursue further development, ever hopeful that the necessary funds would become available.

Sixty percent of expenditures were devoted to the School of Mathematics, with sixty percent of this item allocated to professors' salaries. The investment in human resources had driven the Institute's meteoric rise. The permanent faculty of Einstein, Veblen, Weyl, Alexander, von Neumann, and Morse, together with the visitor Pauli, formed a truly notable intellectual aggregation. Each, excepting Einstein, gave a course or seminar. Everyone interacted with members, the university faculty and graduate students. Discussion and contemplation of deep mathematical ideas permeated Fine Hall. Princeton had achieved the ambience of Göttingen earlier in the century.

The School of Humanistic Studies consisted of just its two professors, Panofsky and Meritt. Meritt was on leave for the year at Oxford and in Athens. As Earle continued to convalesce, the School of Economics and Politics opened with only Mitrany and Riefler in residence. The personnel levels of the new schools resembled that of Mathematics three years earlier when its roster was comprised of Einstein, Mayer, and Veblen, with only Veblen in Princeton. However, the circumstances then were vastly different. Resources were available to finance much of the expansion conceived by Veblen.

Income for 1935–1936 was projected at $300,000, leaving a $50,000 surplus. Even the latter figure was deceptive, in that it failed to account for the Institute's pension obligations. One of Flexner's earliest objectives was to ensure a comfortable retirement for his faculty. The agreements with Einstein and Veblen included an $8,000 pension, with lesser sums for surviving widows. To implement these provisions Flexner subsequently arranged for joint contributions to the program at the Teachers Insurance and Annuity Association of America (TIAA). The difficulty was that neither Veblen nor Einstein had sufficient time, prior to retirement, to accumulate anywhere near the $8,000 guarantee. While later faculty offers merely contained the TIAA arrangement, the trustees had promised $8,000 to Flexner as well. This meant that the Institute might soon be obligated to underwrite retirement payments of magnitude approaching one half its surplus.

Flexner either ignored or failed to understand the pension accounting. He saw a $50,000 surplus that required careful management until supplements became available from the founders or foundations. Every move was calculated in terms of winning Bamberger's favor. Flexner was familiar with how Harper's machinations at Chicago had alienated Rockefeller. Rather than seeking bailouts for unfunded commitments, Flexner presented opportunities to which he hoped Bamberger would willingly subscribe.

The campaign was two-pronged. Every compliment directed at the Institute was rapidly transmitted to Bamberger and Fuld. Pitches for new personnel and initiatives were prefaced or suffixed with the hint "should funds become available." Flexner thought he was being subtle, but the propaganda bombardment would have been transparent to a less astute person than Bamberger.

The wish list was substantial. Flexner's top priorities were to hire Lowe and to launch Riefler's research program. Meanwhile it was becoming apparent that infrastructure could not be deferred any longer. Faculty needed offices. Morse and Pauli were absorbed into Fine Hall, and Morey made space for Panofsky in the art and archeology building. No suitable accommodations existed for Riefler and Mitrany. Flexner could only propose that the social scientists share his rented quarters in downtown Princeton or work at home or in the university library. Matters would soon get worse when Meritt and Earle took up residence. There was a pressing need for the Institute to acquire its own property.

Flexner was finally facing the fiscal challenges that typically plagued university administrators. He had had remarkably good fortune during the first half of the thirties. Most university presidents entered these depression years with long-standing obligations that were barely met by declining revenues. While Flexner criticized their profligate spending, he enjoyed the unique position of operating without prior encumbrances, with a large endowment, and a free hand. Now the income was virtually exhausted. Riefler's program alone could deplete all of the surplus. Land and a building were far more expensive. Reaching into the endowment would necessitate retrenchment.

If the Institute were to maintain any momentum, it was imperative for Bamberger to step up and pay for the campus. The correspondence between Bamberger and Flexner seems to indicate that some assurances, short of a blank check, were given along these lines. Selection of a site was tasked to the Committee on Buildings and Grounds, consisting of Maass, Veblen, and Aydelotte, with Flexner and the founders holding ex-officio status.

There was agreement that the Institute should procure a tract contiguous with the substantial property holdings of Princeton University. The strategy was designed to solidify their relations and enhance further cooperation. One scheme under serious consideration was to engineer a land swap permitting the Institute to erect a building on a site currently owned by the university.

After considerable study Flexner and the committee decided on the Olden Farm as the best available location for the Institute campus. Together with some adjacent parcels the site comprised roughly 200 acres at a cost of a quarter million dollars. Flexner found himself conflicted. The acquisition of so much land was a clear violation of the sacred Gilman precept "men over buildings." Yet Flexner felt that the large tract was a prudent investment. He knew that other universities had made short-sighted decisions to economize on land purchases. In later years more space was needed, and the surrounding area was unobtainable or it was developed in an undesirable manner.

Weighing the logic of sound investment against long-held doctrine omitted one of the most important factors from the calculation. The conservative Bamberger, who was expected to write a large check, might deem the plan to be too extravagant. After all, his own estate had been available for free. Flexner maneuvered for a position to give himself deniability while the Institute acquired the land. Writing to Bamberger, Flexner expressed his own reservations over the magnitude of the purchase, attributing the outcome to the overzealousness of Maass and Veblen.[2]

Bamberger had little interest in the assignment of blame. He was disturbed both by the property transaction and what he viewed as an ill-advised School of Economics. The Institute, under Flexner's guidance, now appeared headed off track. Bamberger replied as follows:

Dear Dr. Flexner:

Your letter of October 28 was quite impressive, as it expressed the thought that possibly some of our coworkers in the management of the Institute were inclined to rush along with more haste than wisdom. Mrs. Fuld has repeatedly commented on a policy of acquiring so much land for an institution that proclaimed not size but highest standards. This also has been my feeling.

After our present commitments are completed, our resources will not permit of further expansion at the present time. So far everything has developed beyond our fondest expectations, thanks to you. Nor have I misgivings about the future.

Kind greetings to Mrs. Flexner.[3]

Flexner, the master at channeling vast sums from philanthropists to worthy projects, had mishandled his own benefactor. It was unclear what was included in the "present commitments," but Bamberger had served notice that he was suspending support for the Institute. Bamberger's fortune had been Flexner's to lose, and it appeared that he had lost it.

One of Flexner's most remarkable qualities was his capacity to take a punch. Bamberger's decision, like the earlier actions of Birkhoff, could have devastated Flexner's aspirations for the Institute. Again, he reacted without any discernible bitterness. First, Flexner graciously acknowledged the letter from Bamberger. Then Flexner resumed his efforts to lift the two new schools to the level of Mathematics, albeit on a more frugal basis.[4]

The most expansive plans were for the School of Humanistic Studies, where Morey continued to exert considerable influence. Like Veblen, Morey possessed impeccable taste and stellar connections in his field. The difference for Flexner in his two consultants was that Veblen's interest was vested in the Institute while Morey's was in the university. Bringing distinguished scholars to either place was an undeniable benefit to the other. Under a tight budget, however, a limited number of initiatives could be pursued. Priorities for the Institute did not necessarily match up with those for the university. These differences tended to emerge with the selection of temporary appointments rather than the permanent faculty.

One distinction between the nature of research projects in archeology and in mathematics was that prosecution of the former often required large groups. For example, several Princeton University faculty were involved in the excavation of Antioch. Carrying out the myriad aspects of field work and publication entailed a number of associates and assistants. University budgets and foundation grants rarely provided the flexibility and funds to cover all of the needed staff. Morey was attracted to the Institute as a resource that he understandably perceived as possessing deep pockets.

Morey wasted little time in carving out his own entitlement. Serious collaboration with Flexner began in March 1934. While it would be another year until the hiring of Meritt and Panofsky, the Institute's first Humanistic Studies outlay was made for the 1934-1935 session. Flexner agreed to a one year grant of $6,000 in support of two scholars working on projects at the university. Flexner first announced the expenditure to his trustees at the April 1935 meeting, requesting a one-year extension and disingenuously portraying the purpose as "making a survey of the resources in the field of humanism from New York to Washington."[5] Such a survey may have been undertaken, but the project's real objective was to support work already begun on Visigothic architecture in Spain and on Old Testament illustration.[6]

The $6,000 was a significant commitment of Institute resources. What Flexner intended as a short-term stipend to help out the university, the trustees were led to believe was seed money for the School of Humanistic Studies. Morey quickly came to regard the $6,000 as a cushion for his operating budget. He programmed the money for Antioch workers and other projects as needed, giving no consideration to the Institute. It is unlikely that Morey knew, or that Flexner volunteered, the precarious state of Institute finances. Morey was hoping that the subsidies would continue and expand. If Flexner acquiesced, development of the new schools faced a further challenge.[7]

Viewing the payment as a retainer for Morey's advice on faculty selection, Flexner received a solid return. Morey's original blueprint called for Panofsky, Lowe, Dinsmoor, Meritt, and an Islamist to be named later. By the fall of 1935 Panofsky and Meritt were on board, Lowe was the next priority, and Morey had identified his Islamic archeologist. He was another German Jewish refugee, Ernst Herzfeld.[8]

Born in 1879, Herzfeld received a PhD from the University of Berlin in 1907. His dissertation was on Pasargadae, the ancient capitol of Persia (located in what is currently Iran). Four years later Herzfeld led a pioneering excavation of Samarra in north central Iraq. Over the next quarter century Herzfeld participated in a large number of Near Eastern expeditions. Aided by an indispensable mastery of local languages, Herzfeld established himself as the pre-eminent authority on Islamic archeology and art.[9]

In 1931 Herzfeld began the excavation of the ancient Persepolis palace in Iran. At this time he was associated with the University of Berlin and the Oriental Institute of the University of Chicago. Herzfeld left the Persepolis project in 1934, ending his connection with Chicago. It was not a good time to return to his position in Berlin. Over the next year Herzfeld severed his relations with Germany. When the well-connected Morey learned that Herzfeld was a free agent, the nomination was forwarded to Flexner.

At the age of 56 Herzfeld's career was entering its final phase. He had spent decades in the field, making seminal discoveries and accumulating an extensive personal collection of artifacts. What he now needed was an opportunity to reflect and write up his work for publication. Morey was also anxious for Herzfeld to train some young Princeton scholars in a specialty for which little expertise was available in America. Herzfeld's combination of research standing and teaching potential made an ideal profile for Flexner.

Flexner was faithfully following Morey's plan for the faculty in Humanistic Studies. Adding Lowe and Herzfeld to Panofsky and Meritt would constitute a worthy collection of scholars within a niche of archeology and art history. It is unclear if the archeologist Dinsmoor was ever pursued, or whether it was determined that he was immovable from Columbia. In any event, Morey began pushing another appointment that was of questionable wisdom for the Institute.[10]

The excavation at Antioch was winding up a five-year cycle. Arrangements for a long term follow-up were under negotiation. Among the current staff was W. A. Campbell, who split the year between his duties as field archeologist at Antioch and an associate professorship at Wellesley. With Campbell in line for a more desirable position at Wellesley, Morey was seeking the means to lock up his archeologist for the sequel project at Antioch. A professorship at the Institute would free Campbell to work on the excavation and spend the remainder of each year analyzing the results in Princeton. It is puzzling that Flexner gave serious consideration to the proposal. Campbell was a young scholar, whose research accomplishments fell far short of von Neumann and Meritt. Morey's scheme accrued its benefits to the university while the Institute absorbed all of the cost. Flexner, however, was anxious to accommodate Morey and the university.[11]

Over November 1935 Flexner pondered possible initiatives to pursue the following year. It was his intention to request approvals from the Executive Committee at its meeting scheduled for the first week in December. In addition to the positions for Lowe, Herzfeld, and Campbell, Morey was seeking to regularize and enlarge the Institute's funding to the university archeology department. Flexner had other considerations as well. He was hoping to advance the School of Economics and Politics, and, in particular, Riefler.

Riefler's research methods relied heavily on statistical analysis. In accepting the Institute professorship, Riefler received Flexner's assurance that statistical staff and resources would be forthcoming. In October, Riefler outlined a number of problems that he hoped to examine. Durable goods was not among them. The program was in the area of finance with emphasis on the recent crisis in money and banking. The research design was to bring in senior visitors to collaborate for periods of about a year. Since regular salaries of these individuals were likely to be considerably higher than those of members in mathematics, Riefler planned to solicit these funds from outside foundations. He made the reasonable request that the Institute furnish offices and provide secretarial and statistical support. Flexner responded favorably. He again assured Riefler that the money would be available, merely advising that the problems might best be attacked serially rather than in parallel.[12]

There were other needs easily foreseeable over the next few years. These included a new building, economics faculty, and members for the new schools. The existing $50,000 surplus and Bamberger's final endowment contribution were inadequate to sustain this development. Flexner needed a new source of revenue. For years he had anticipated contributions from the Rockefeller charities and other foundations. Although not one cent had ever been realized, Flexner renewed his efforts. The yield was modest.

For Herzfeld, Flexner obtained a once-renewable salary supplement of $2,000 from the Emergency Committee in Aid of Displaced German Scholars. Lowe was in the midst of an ambitious project to produce a ten-volume photographic catalog of all Latin literary manuscripts prior to the ninth century. He had received substantial support from a five-year Rockefeller Foundation grant. With the Rockefeller grant exhausted, Lowe was relying on a $4,500 stipend from the Carnegie Institute that covered his research expenses, but not salary. Flexner obtained assurance from the Carnegie Institute of a multiyear continuance of their subvention. The bottom line was that Flexner had raised a total of $4,000 toward the lifetime salaries of his Humanistic Studies faculty.[13]

Flexner approached the December 6, 1935, Executive Committee meeting with "the hope then to be able to take a further step in each of the two new schools."[14] For Economics, Flexner came away with nothing. The Executive Committee did not authorize any of the statistical resources that Flexner had promised to Riefler. While no minutes of the meeting were distributed, this breech of faith can only be explained by Bamberger asserting a veto over further support of economics.

The Executive Committee did approve expansion in the Humanistic Studies faculty, provided it could be accomplished within a moderate increase in expenditures. Flexner moved immediately on Lowe, securing him with the $10,000 salary that they had previously discussed. After acceding to Morey's recommendation of $6,000 for Campbell, there was little money left for Herzfeld. The terms of the offer to Herzfeld signified a dramatic shift in Flexner's overall approach to being chief executive of the Institute. Facing severe fiscal pressures, pragmatism and thrift suddenly trumped idealism.[15]

Salary offers in the "paradise for scholars" had evolved from the gargantuan $20,000 for Birkhoff to the large increases for Veblen and Morse to the matching figure for Panofsky. Herzfeld's compensation was set at $6,000, significantly below the level commensurate with his qualifications. What was especially striking was the manner in which the offer was packaged, combining $2,000 each from the Emergency Committee, the Institute, and New York University. Herzfeld was not hired on a full-time basis. He would teach one day each week in New York.[16]

Flexner was compromising two of his most fundamental principles. These were that Institute faculty receive generous remuneration and engage in full-time service. Both of these tenets were Flexnerian in origin, rather than derivative of Gilman. The betrayal would have been less notable if Flexner himself had not been such a strident critic of university presidents adapting to their own exigencies. Flexner was nearing the zero sum circumstances of his peers. To make new hires required some combination of exploitation, deficit financing, and cuts to other programs.

Bamberger's 1936 contribution to endowment came in at just under one million dollars. Nearly $300,000 was needed to cover the land purchase which had reached 265 acres. Located on one of the parcels was a building that, with modification, could be adapted for office use by Institute faculty. There was less urgency to construct a new building, but some locus for Institute activities would be needed in the not too distant future.[17]

The endowment increase yielded more than enough boost in income to cover the new faculty. Budget arithmetic, however, is never simple, and the appointments had implications that went beyond the first year salary. The $2,000 grant from the Emergency Committee was only for two years. The Institute's contribution would double in Herzfeld's third year. More-over, Flexner's offer letter to Herzfeld gave "the assurance that as resources of the Institute for Advanced Study increase, as we have every reason to believe they will increase, your salary will be raised."[18] To Lowe, Flexner agreed that the Institute would indemnify him against severance of the Carnegie grant. Flexner was moving toward the tactics of Harper, stipulating future considerations that were predicated on, at best, questionable budget forecasts. Failure to realize these emoluments was likely to lead to bad feeling. A more systemic problem was that, with fewer than ten years to retirement, Herzfeld and Lowe would accumulate meager pensions.[19]

An immediate opportunity to enhance the salary of Herzfeld arose when the Campbell appointment fell through. Flexner and Morey had miscommunicated over how the offer was to be handled. The result was that Campbell, as well as the Wellesley administration, learned of his Institute appointment when a notice appeared in the *New York Times*. The premature press announcement was embarrassing for all of the parties. The Institute decided, at least temporarily, to yield to Wellesley on Campbell. The $6,000 windfall did not benefit Herzfeld in any way. Morey reprogrammed the money to meet his needs at Antioch and elsewhere. Later, when the Antioch concession was renewed, Campbell received a half-year appointment at the Institute with a more fitting title. Rather than a professor, he became the Institute field archeologist. For half of each year, through the term of the project, Campbell performed field work for the university.[20]

As Flexner scrimped to appoint Herzfeld, he sought to reclaim money from the Mathematics budget. There were two lines that Flexner had repeatedly pointed to as subject to review in the event of financial pressure. These were the visiting professorship and the large allocation for member stipends. Veblen and the mathematicians could defend both on their merits. Dirac and Pauli had brought distinction to the Institute while the members were a creative, cost-effective means of advancing the Institute's mission.

The Veblen-Flexner relationship, so successful in prosperous times, was facing the strains of financial challenges. As Flexner sought economies, Veblen was determined to obtain new funding. Three of Europe's leading mathematicians were making visits to Harvard to participate in its tercentenary. Veblen urged that each be invited for extended follow-up visits to Princeton. In addition to this one time outlay, Veblen continued his campaign to equalize all mathematics faculty salaries at the $15,000 level enjoyed by himself, Weyl, and Einstein.

The outcome was mixed. Flexner succeeded in permanently scrapping the visiting professorship. Veblen obtained $6,000 for the Europeans. Flexner compromised on salary, raising Alexander and von Neumann to the $12,500 figure of Morse, higher than that of any professor in the other schools. On the member budget, Flexner attempted to impound uncommitted funds. He was outmaneuvered by Veblen, who cited prior informal arrangements. The overall effect on the mathematics budget was more or less revenue neutral. The partnership between Flexner and Veblen, however, was headed for trouble.[21]

At this point Flexner's only bitter faculty relationship was with Mitrany. The differences that had arisen in their correspondence were intensified in personal meetings at the Institute. When Mitrany requested a university instructor to serve him as a part-time assistant, Flexner refused. Mitrany then vented, alleging that Riefler and the mathematicians were receiving preferential treatment. An offended Flexner concluded that Mitrany was an ungrateful whiner. Regretting that he had ever made the appointment, Flexner suggested that Mitrany might be happier at another institution. Flexner was hoping that Mitrany would resign. Tenure precluded more extreme measures.[22]

Flexner was concerned about the School of Economics and Politics. There seemed little that he could do to advance it. Earle continued to make slow progress in his recovery, battling occasional setbacks. He was able to visit the Institute during the spring of 1936, but permanent residency was continually moved back to some indefinite point in the near future. Flexner remained unfailingly supportive and patient, never pushing Earle to help the flagging school. Flexner's immediate hope was for Riefler to emerge as a star. The frustration was that the institutional support needed by Riefler, so modest by the scale of Flexner's past enterprises, was totally out of reach.[23]

It appears that Flexner was less than candid with Riefler over the funding problems. For Riefler attempted to refine his vision for the Institute economics school, maintaining the budget at an unrealistic level. In a March 1936 Riefler memo the subject was a rhetorical question: "Should the Institute concentrate its work in economics in the field of Finance?"[24] Riefler answers affirmatively, setting out a program in which the Institute "assume(s) leadership in formulating a broad inquiry into the causes and phenomena of the financial crisis." The proposed study of the financial aspects of the depression was to include participation by the Federal Reserve, government agencies, the Rockefeller Foundation, and universities. The scope and partnerships were what Flexner originally had in mind for the school.

Despite its appealing features, Flexner realized that Riefler's plan posed insurmountable difficulties to its implementation. The budget called for $50,000 in the first year and $100,000 thereafter. The money was simply unavailable, even in Flexner's rosy dreams. Not only was Bamberger likely to be hostile, but Stewart had provided Riefler with a lukewarm reaction. As always, it is difficult to interpret Stewart's actions. He was contemplating joining the faculty, and may have wished to keep the School's agenda open for his own design.[25]

This time, Riefler's proposal was not even presented to the Board. Flexner did manage to come up with some support for the scholar he regarded as so promising. The consolation prize was a $1,500 stipend to fund a summer trip to Europe. Flexner was hoping that consultations with economists in England and France would lead Riefler to a cheaper research methodology.[26]

The new austerity at the Institute did not prevent Flexner from considering a major capital outlay. In 1936 he learned that a large collection of Chinese books was on the market. An American, G. M. Gest, had accumulated over 100,000 volumes during his lifetime. The Gest Library was an extraordinary resource for scholars in Oriental Studies. When Gest found himself in severe economic difficulty, he began seeking a buyer for the collection. The Library of Congress and a number of universities coveted the Gest Library, but the $125,000 price tag was beyond their reach on short notice at that time.[27]

Gest was desperate for funds. If no buyer for the library were found, Flexner feared that the collection would be dispersed. Parts would be sold or auctioned to various institutions and individual buyers. With the collection scattered around the world, its usefulness to scholars would be forever diminished. When Flexner served on the General Education Board he relished his role in making interventions to prevent such tragedies. The Gest Library now needed a home. Although neither the university nor the Institute was presently engaged in research that utilized the materials, Flexner viewed Far Eastern studies as an area of rising importance. He reasoned that locating the Gest Library in Princeton would position both institutions for future entry into the field.[28]

Flexner persuaded the Rockefeller Foundation to match the Institute's $62,500 in purchasing the Gest Library. The Institute became the owner, but the Rockefeller contribution was conditioned on the university holding perpetual access to the collection in Princeton. It was no problem for Flexner that the terms were overwhelmingly in favor of the university. He was happy for an opportunity to give something back for all that the Institute had received. Flexner portrayed the purchase to his trustees as an Institute supplement to the library resources that were so generously provided by the university.

He failed to mention that the expense of owning the Gest Library went beyond the purchase price. The books could not simply be stored in a large closet and held until professors were appointed. The collection required the regular attention of a curator. Together with other costs, the library would become a $6,500 annual expense. Since divestiture required the consent of the university, the maintenance burden was forever on the Institute. Flexner never doubted the wisdom of the purchase. He did not understand that rescuing the collection was the appropriate role for the Rockefeller charities, but an unaffordable luxury for an institution with pressing needs.[29]

The summer 1936 acquisition of the Gest Library concluded a remarkable first year for the School of Humanistic Studies. Flexner would modestly tell the Board that the School had "made a definite beginning."[30] It was much more. The important art historian Panofsky was permanently settled in Princeton. Meritt made such a strong impression at Oxford that he was awarded an honorary degree, quite an accomplishment for a classical scholar still in his mid-thirties. The experts Lowe and Herzfeld were appointed and were bringing their own magnificent libraries and collections to Princeton. Finally, Flexner had begun the process of hiring an outstanding archeologist who would become the fifth professor in Humanistic Studies, and the first woman on the Institute faculty.[31]

Hetty Goldman was born in 1881 into a prominent German-American Jewish family. Her grandfather was a founder of the Goldman Sachs investment firm and her father, Julius Goldman, was an educated and wealthy New York attorney. Hetty attended one of the best private schools in New York, the Sachs School for Girls. It was run by her uncle, who himself held an interest in archeology. In 1903 she completed her undergraduate work at Bryn Mawr College with a double major in Greek and English. Graduate study at Columbia and Radcliffe led to a master's degree in 1910 and an article dealing with mythological illustration on Greek vases. These credentials positioned Goldman to become the first woman to receive a prestigious Harvard fellowship for study at the American School of Classical Studies at Athens.[32]

It was in Greece that Goldman found her true calling. She was assigned to her first excavation, at Helae on the east coast. The research earned her a PhD from Radcliffe in 1916. Over the next several decades, interrupted from time to time by wars, Goldman led archeological expeditions to sites in Greece and Turkey. Her work brought important new understanding about the people in this region during the classical period and earlier. Funding for Goldman's work came from the Fogg Museum of Harvard, Bryn Mawr, and her father. In 1934 she began work on an excavation at Tarsus, on the southeast coast of Turkey. Goldman was training a number of Bryn Mawr students as apprentices in the field, but she did not hold a position with traditional academic duties. Like Herzfeld, Goldman had reached a career point where she needed an opportunity to convert her research studies into publishable form.

The appointment of Goldman in 1936 to the School of Humanistic Studies appears to provide evidence of the Institute's faithful adherence to its nondiscriminatory founding ideals. One might expect that Flexner learned of Goldman through her professional reputation, and then solicited opinions of faculty members in the School of Humanistic Studies. Continuing this hypothetical scenario, Flexner would have received glowing reports that prompted the appointment of Goldman at a salary commensurate with her colleagues. These were not the circumstances surrounding the Goldman offer.

It was through family connections that Flexner became acquainted with Goldman. Early in the century, Flexner came into contact with Goldman's uncle through their common interest in education. The families began to socialize, leading to a long-term friendship between Hetty's father and Flexner. After completing the appointments of Lowe and Herzfeld, Flexner began soliciting references on a position for Hetty. The knowledgeable Morey deferred to Meritt. Flexner wrote Meritt at Oxford in February 1936 to inquire whether Goldman's scholarship and archeological work were of suitable quality. Meritt was in no rush to bring Goldman onto the faculty. He put off his response until the summer, when there would be an opportunity for discussion. After meeting with Flexner, Meritt wrote a letter of support. The appointment of Goldman was approved at the October 1936 trustees meeting. Goldman's salary was only $200 a month, less than half that of the underpaid Herzfeld.[33]

Flexner presented the Goldman nomination to the Board in a manner that concealed the introduction of such a large disparity in faculty salaries. His tactics were clever. The main event at every meeting was the delivery of the Director's report. At this time Flexner would review the (excellent) state of the Institute and provide his rationale for the actions he planned to present for approval. At the end of the meeting, Flexner's administrative assistant read the formal motions as they were submitted for vote.

Two other staff appointments were proposed at the October meeting. These were the Gest Library curator and Edward Capps, a classics professor who had recently retired from the Princeton University faculty. In his report, Flexner pointed to Capps' vigor and to the Institute's opportunity to gain the services of a leading Hellenist who would collaborate with Meritt. Flexner proposed that Capps be compensated with a monthly supplement to his university pension. "The arrangement would be made on an annual basis so that the obligation that I am recommending would terminate unless annually renewed."[34]

The Director's next sentence introduced Goldman: "I propose also to take this same action in reference to a splendid young woman who has for some years been engaged in Grecian and Asiatic excavations." The follow-up explained that Goldman's work complemented that of Meritt and Capps. With no mention of Goldman's title and terms, the clear implication was that her appointment was also to be on a yearly basis.

When the nominations were presented, Capps was first. The particulars were a visiting professorship for one year at $200 per month. The wording for Goldman was similar, except for deletion of the adjective "visiting" and the phrase "for one year." A trustee who had followed the Director's Report, and not carefully parsed the nomination, was likely to leave with the impression that both Capps and Goldman were visiting professors.

Veblen was surprised to learn afterward that Goldman's appointment was permanent. Maass, who was dubious of the Gest acquisition, reacted similarly to the arrangements for the curator. Flexner brushed off both objections. An imperial director could get away with making appointments in this manner, but it was unwise to leave Veblen and Maass with the feeling of having been railroaded.[35]

Goldman herself was thrilled with the appointment. As for the token salary, Flexner explained that it was a temporary arrangement for which an adjustment could be expected in the future. Since Goldman's appointment began in November, after the term had started, she inferred that the raise would be instated for the following academic year.[36] When the next session did begin, the salaries of Riefler and Meritt were increased to $15,000 while Goldman remained at $2,400. As the years went by without any change, Goldman seethed.*

Flexner's exploitation of the Goldman family took place on both sides of the budget ledger. When Hetty decided to resume her field work at Tarsus, her father came forward to provide most of the funds. Julius Goldman requested that Flexner treat the contribution as if it were anonymous. The donation was laundered so that it appeared that the Tarsus expedition was being supported by the Institute. At the next Board meeting Flexner announced the contribution to Goldman's work, adding that the anonymous donor was being cultivated to take a larger part in development of the Institute. Not only was a woman engaged to work with meager compensation, but Flexner was actually looking to realize a profit for the Institute.[37]

Goldman joined Herzfeld, Lowe, Meritt, and Panofsky to form an impressive faculty for the School of Humanistic Studies. There were two arche-

---

*In 1945, two years prior to her retirement, Goldman's salary was increased from $2,400 to $10,000.

ologists, a paleographer, an epigrapher, and an art historian. Each was an outstanding scholar who was devoted to research. Flexner had done extremely
well on a considerably smaller budget than was invested in the mathematics faculty. Yet Humanistic Studies operated with two distinct disadvantages
compared to its eldest sibling. They had neither Fine Hall nor a $40,000
budget for members and assistants.

Goldman and Meritt moved in with Riefler and Mitrany to the building recently acquired by the Institute. Lowe elected to work at home, as did
Herzfeld, who declined space offered by Morey in the university building
where Panofsky was located. With Goldman frequently abroad, faculty interaction was limited. Isolation was not a problem for the professors, who were
each devoted to their own project. However, it was difficult for the School of
Humanistic Studies to create a physical seat of learning such as was achieved
in Fine Hall by Mathematics.[38]

Flexner loved to wander around Fine Hall, overhearing informal conversations about deep mathematical ideas. Although he never understood the
substance Flexner was thrilled to witness the intellectual activity. The objective of the Institute was to elevate research in America. Only so much could
be accomplished by faculty publication. To succeed the Institute needed to
prepare the next generation of scholars. Flexner could see this happening
among the members in Fine Hall.

Each year Veblen and his colleagues selected about 40 members and assistants for the School of Mathematics. Some came to collaborate with professors. Most were promising or established scholars whose projects were independent of the faculty. The $40,000 went a long way as Princeton became
a magnet for recipients of national and international fellowships. The attraction was simple. Beginning with its excellent library, Fine Hall provided an
extraordinary environment for mathematical research. There was the constant stimulation of interacting with others who shared an enthusiasm and
understanding of mathematics. Every day legendary mathematicians lectured
on cutting-edge developments. Finally, the Institute offered complete freedom to think and work. No mathematician could ask for more ideal research
conditions. The impact of the year-long memberships carried over to future
research careers and their host campuses.

Flexner was hopeful of spreading the member paradigm to the other
schools. By 1936 Humanistic Studies, with its faculty in place, was positioned
to embark on a similar program. The obstruction appeared to be money. For
1936-1937 $10,000 was budgeted for stipends in Humanistic Studies, virtu-

ally all of it for Antioch and other university projects. Flexner wrote to the Carnegie Corporation requesting a three-year bridge grant of $25,000 per year for Humanistic Studies. As was becoming his habit, Flexner offered the baseless assurance that "it would be beyond question capitalized by friends of the Institute by the end of the period."[39]

Although limited funds were provided by the Carnegie Corporation, Humanistic Studies would never replicate the program for members in mathematics. One factor was the differing nature of the disciplines. For example, mathematics was considerably more portable. However, the overriding reason for the success in mathematics was Veblen. Veblen designed the postdoctoral program, implemented it, and fought each year to maintain it. His efforts included annual canvassing for new members and defending against budget cuts by Flexner and the Board. Whatever their differences, Flexner and Veblen shared a fundamental commitment to the development of young scholars.

There was no one in Humanistic Studies to assume Veblen's role. The faculty were focused on their own research. Morey was the real force in shaping the school. Shortly after Goldman was appointed, Morey compiled a wish list for Flexner. The primary items were to continue support of Antioch personnel, provide release time to Princeton faculty, and create new professorships. Morey's vision for the Institute was limited to its role in advancing the program at the university. Even if $40,000 and a building such as Fine Hall had been available in Humanistic Studies, no appointments comparable to Gödel would have resulted.[40]

Humanistic Studies had not reached Mathematics, but it had made enormous strides in its first two years of operation. Flexner clearly identified his next tasks. The Institute needed to construct a building and to strengthen the program in economics. Since the enhancement of Humanistic Studies was achieved within the income from Bamberger's last endowment contribution, the Institute's surplus remained at $50,000 over 1936-1937. The figure was subject to constant pressure. With fourteen professors, minor unanticipated expenses would continually arise. Meanwhile income was beginning to decline and there was little hope of any intervention by Bamberger. Even if no professors were added, the surplus was likely to diminish.

While Flexner struggled with the budget challenges, the Mathematics faculty was assessing its own needs. The time seemed ripe to add strength in theoretical physics. There was reason to believe that some of Europe's greatest scientists would accept a call to the Institute. Einstein advocated Schrödinger, whose Princeton baggage made him a non-starter with Flexner. It was hard to find fault with the other nominees, Dirac and Niels Bohr. Both were Nobel Laureates. Dirac, still in his mid-thirties, had been pushed by Veblen from the start. The 51-year-old Bohr was a pioneering figure in discerning the structure of atoms.[41]

Despite his earlier reluctance to hire Europeans, Flexner was receptive. If Bohr and Dirac joined Einstein, the Institute would immediately become the world's leading center for theoretical physics. Flexner liked the scenario of members moving on to American universities, elevating science throughout the country. The obstruction was the intractable budget situation. Flexner was unwilling to commit any additional income to the School of Mathematics. To bring in the theoretical physicists, he would need to obtain a new funding source.

In March 1937 Flexner applied to the Rockefeller Foundation for a three-year grant to support the salaries of Bohr and Dirac. The petition echoed the unrealistic budget forecast that Flexner had pitched to the Carnegie Corporation: "At the end of a period of three years the resources of the Institute will have so increased that I feel certain it will not be necessary to call on the foundations for further support for these men."[42] The hollow assurance was not tested. The Rockefeller Foundation declined the request, explaining that it had already provided substantial funding for the programs of Bohr and Dirac at their institutions in Copenhagen and in Cambridge. It made no sense to participate in luring them away from the Rockefeller infrastructure.[43]

Theoretical physics remained on Flexner's to do list, but it was not the highest priority. Until there was a change in Bamberger's attitude, a building was out of the question. The greatest urgency for Flexner was to do something for the School of Economics and Politics. The school's first two years had been disappointing. Mitrany was in the doghouse and Earle was sick. Perhaps most frustrating for Flexner was that Riefler seemed to be reverting to his bureaucratic past. He had taken on consultantships with the League of Nations and the Department of the Treasury.[44]

Riefler's actions were understandable. The Institute's failure to establish a statistical staff had been a major setback for his research. Riefler adapted to these circumstances by transferring his financial program to the auspices of the National Bureau of Economic Research. With obligations that included management and consulting at other locales, Riefler was not a full-time Institute research economist.[45]

Flexner wanted to redirect Riefler's efforts. He earmarked the evaporating surplus for faculty in economics. If a collaborator were brought in, Flexner

reasoned that Riefler's research would flourish. The difficulty came in identifying a suitable candidate. The search for economics faculty had been almost
continuous since the early days of Institute planning. In all this time only
Riefler and Stewart had impressed Flexner as viable Institute professors. Both
Riefler and Flexner continually held out hope that Stewart would decide to
join them. Stewart's signals, such as they were, did not discourage. Some
British economists were also under consideration, but there was no name to
present in the foreseeable future.[46]

The uncertainty over economics faculty was just one element of Flexner's
frustration in approaching the summer of 1937. The budget surplus for the
coming year projected to no more than $35,000. Veblen was pressing for
offers to Dirac and Bohr. The Secretary of the Treasury wanted too large
a piece of Riefler. Even the prospects for a building had gone from bad to
worse. A university site was no longer viable.

Nothing was resolved by mid-summer, and the 70-year-old Flexner
headed to Ontario for his summer retreat. In late July he made a one-week
side trip to Murray Bay in Quebec where the founders were vacationing.
Bamberger was 82 and Fuld was in poor health. It had been nearly two years
since they had frozen support for the Institute. Flexner desperately needed
additional funds to carry out his plans for development. Future opportunities for personal persuasion were likely to be few, if any. Flexner seized the
moment and presented a strong appeal to the founders for economics, mathematical physics, and a building.[47]

Following their long-established roles, Flexner spoke and Bamberger listened. Flexner hit on his traditional themes of quality and conservatism.
He reiterated that funds should only be used to hire exceptional people. It
happened that rare opportunities were arising for the existing schools. Unfortunately, current income was insufficient to take advantage and act. Flexner
urged an infusion of several million dollars into the endowment. This would
enable him to place the Institute on a high plateau during his remaining years
as Director.[48]

Bamberger still had serious reservations about putting more money into
the Institute. He was also a courteous host with a genuine interest in Flexner's
vision. As the week passed Flexner watched intently for some affirmative
sign from his quiet benefactor. Whatever it was that transpired, Flexner
returned to Ontario with a feeling of exhilaration. Letters to Leidesdorf,
Riefler, and Maass raved about a successful mission. To Riefler, Flexner said
that the founders "told me to go ahead" with the proposed development in
economics.[49] With Leidesdorf Flexner was more circumspect: "While no

promises were made, I have the feeling that I have the authorization to go ahead."[50] To Maass he wrote: "I feel no doubt that, if we can find the persons, there will be no trouble about the funds."[51]

The inconsistencies in the letters raise questions about the communication that took place at Murray Bay. Flexner was certain that the salaries of outstanding new professors would be capitalized. Was this wishful thinking or the intent of the founders? For the October 1937 trustees meeting, Bamberger was present and Fuld was absent. In his report, Flexner rehearsed their Murray Bay discussions regarding new appointments: "Our present funds do not permit expansion to this extent, but after consultation with the Founders I have permission to investigate the possibilities."[52] This statement is probably a fairly complete representation of what took place. The founders did not rule out further exploration, but they made no commitment for support. Bamberger's attitude toward the latter issue would become clear over the next year.

On another development front, Flexner relayed news to the trustees of a major breakthrough. The founders "informed me that they wished to furnish the Institute with funds necessary to erect our first building without drawing upon the capital funds." [53] The building was to be named Fuld Hall, in honor of Mrs. Fuld and her late husband. Flexner would begin work immediately by polling the schools and collating their needs. The trustee Committee on Buildings and Grounds of Maass, Aydelotte, and Veblen were to join in the planning process.

The apparent resolution of the budget crisis was a huge relief for Flexner. He approached the 1937-38 academic year dedicated to lifting the School of Economics and Politics to at least the level of Humanistic Studies. The term began on a promising note. Ed Earle had recovered from his struggle with tuberculosis. Earle's comeback was astonishing. After a decade of incapacitation, he thrust himself into a study of international relations. Flexner's faith and long-shot bet were finally realizing a return.

Earle joined Riefler and Mitrany in the Institute's temporary building. With the new opportunities for interaction among the School's faculty, Flexner hoped to steer Mitrany and Earle toward collaborative economic work with Riefler. The effort to channel Earle was no more successful than with Mitrany. Earle was a serious, independent scholar. In setting his research course he allowed no accommodation for administrative agendas.[54]

Reorganization was just a minor ingredient of Flexner's plans for the School of Economics and Politics. Flexner's real hope was to bring in Walter Stewart and one or two other economists. Everything hinged on whether

Stewart would finally leave his firm to accept an Institute professorship. The courtship had been going on for five years. Flexner felt now, more than ever, that Stewart was on the verge of accepting. As the vigil continued, Flexner was forced to turn his attention back to the budget.[55]

Dividends of Institute stocks were declining and the $35,000 surplus projection appeared overly optimistic. In his summer meeting with the founders, Flexner thought that he had gained control over the budget. Since that time, Bamberger's only contribution was $50,000 toward a building fund. It was enough to commission site studies and architectural work, but inadequate for construction. No additions to endowment had been made. By the January 1938 trustees meeting, Flexner was no longer counting on Bamberger to pay for the new personnel. In his report, Flexner alluded to the gifts already received from Goldman's father and the Rockefeller Foundation (in the purchase of the Gest Library):

> As the work of the Institute becomes known, gifts of this character will increase, but meanwhile I feel that I shall have to devote a considerable part of the remainder of this year to procuring the funds needed to bring the School of Humanistic Studies and the School of Economics and Politics up to the level of the School of Mathematics.[56]

Flexner was worried. Without further assistance from Bamberger, the budget situation reverted to that of the previous summer. The surplus, so important to Flexner, appeared on the brink of becoming a deficit. There were already commitments for the following year, such as the increased payment to Herzfeld. Flexner decided it was time for restraint. With Bohr set to visit for one term, Flexner tabled consideration of any additional steps in theoretical physics. Another cost saving measure involved the budgeting of an upcoming residence by Gödel.[57]

Gödel, who had held two prior memberships, had recently obtained another major result. Von Neumann was anxious to bring Gödel to the Institute again. The School of Mathematics proposed a scheme where his stipend was to be appropriated from a new account. Flexner ruled that Gödel's salary was to be drawn from the members' budget. In the ensuing discussion with Veblen, Flexner learned that Mathematics was also regarding Bohr's salary as outside the members' line. The exchange became testy.[58]

Flexner finally received some good news in March when Stewart indicated a willingness to join the faculty. Flexner was eager to follow up. He looked to Stewart to become the savior of the School of Economics and Politics. Stewart was not yet ready to commit, but he did propose terms for consideration.

They included the addition of two other permanent professors and one visiting professor in economics. Flexner was in the midst of preparing the budget. The actual surplus for 1937–1938 was $16,000. Even without Stewart, cuts were required to obtain balance for the following year. To provide for Stewart and his entourage, the Institute needed an infusion of new resources.[59]

The conundrum was excruciating for Flexner. He simply had to have Stewart. However, after years of lecturing the trustees on the importance of a surplus, Flexner could not suddenly present a budget with a deficit. The only solution was for Bamberger to come to the rescue. Yet Flexner knew that Bamberger was opposed to pursuing economics. There was no alternative. Flexner had to attempt something he dreaded. He had to directly solicit the founders to provide funds for a move that Bamberger considered to be ill-advised.

Flexner went to Bamberger, and the outcome was disastrous. After Bamberger refused to back the economics initiative, Flexner fell back on his old, less aggressive techniques. He followed up by asking Bamberger to read a Rockefeller Foundation report on the urgency of making advances in the social sciences. Bamberger's response was to suggest that Flexner obtain the money from the Rockefeller Foundation.[60]

Of course if there were any prospect of adding Rockefeller money to the endowment, no prompting would have been necessary. Flexner was left with a lot of explaining for the Board. Neither Stewart nor the founders were present at the April 1938 trustees meeting. Flexner prepared the ground for a shift in doctrine;

> Little did we think when we made our beginning in 1933 that five years thereafter financial conditions would be as bad as or worse than they were. On the other hand, it is much easier for an old, established institution to slow down its pace than for a young and lusty infant such as the Institute for Advanced Study. While, therefore, I believe that we must proceed with the utmost caution, I also believe that we must proceed with courage and with faith in the future . . . . it is our duty as Trustees to go forward, not backward.[61]

Flexner went on to recall that the Institute had achieved its success by seizing opportunities to hire people of "genius, of unusual talent, and of high devotion, who may be willing to be associated with us." He then came to the point:

> we can in the near future probably associate with ourselves in the department of economics two or three men who correspond to this description. I

am not prepared to state the absolute amount which will be needed to secure them, nor can I state that we will secure them, but the amount will not,
at first, in my judgment, exceed annually $50,000 or $60,000 in addition
to our present expenses. I propose, therefore, to ask the Board to authorize
me to take steps as may be necessary to place the department of economics
upon a basis approximately equivalent to that of the other two departments.
That will involve getting the men and raising the funds.[62]

After considerable discussion Flexner's proposal was approved. The 1938-
1939 budget did not include the possible enhancements in economics. Balance was achieved, but the income assumptions were sufficiently speculative
that the Board included procedures to cope with the contingency of a deficit.
Such an event called for compensation in the following year.

Stewart continued to ponder his decision as he embarked on a trip to Europe at the beginning of the 1938 summer. He found ample opportunities to
explore the possibilities. Accompanying him were Riefler and two economists
under consideration for the faculty, Columbia labor economist Leo Wolman
and Stewart's business associate Robert Warren. In England they intersected
with Flexner and a third candidate, Bank of England advisor Henry Clay.
When Flexner returned to the States in June, Clay, at least, was holding an
offer. The entire plan in economics, however, remained hostage to the brood-
ing Stewart, who was still in England.[63]

While Flexner was crossing the Atlantic, Stewart sent the following cable
from London: "My answer is yes—wholeheartedly and without reservation,
and I am delighted to have made the decision."[64] Flexner never received the
telegram. He remained in suspense until Stewart followed up after his arrival
in the States.

A jubilant Flexner sent an emotional letter to Bamberger that began: "I
was thrilled yesterday as I have not been thrilled and excited since the day
when Professor Einstein accepted a professorship in the Institute, for I had
a belated cable from Mr. Stewart, reading as follows . . . ."[65] Flexner then
alluded to Bamberger's skepticism, conceding that the subject of economics
was in a primitive state. Yet Flexner went on to argue that subjects developed
imperceptibly over long periods of time until, suddenly, a magnificent discovery was made. In physics there were the theories of Newton and Einstein.
From the experience of his own life, Flexner pointed to the eradication of yellow fever. Such breakthroughs were rare, and the risks of pursuing them were
great. "Psychology and economics are harder, but our Institute was founded
for the purpose of attacking the difficult and even the apparently impossible."

It was vital to understand the underlying causes of economic problems and then to work out solutions. It may takes years or even centuries.

Flexner repeated his gratitude for the founders' previous support, and then he begged:

> Now, in probably the last field of study which I shall initiate, I need your faith once more . . . . My one hope is that the three of us with whom the responsibility rests may stand together and use the same faith and the same cooperation and the same spirit in this new field. After long years of waiting we have secured a leader who is universally regarded by those most competent to judge as the ablest person in either Europe or America.

Flexner concluded,

> I realize I am no longer young, and it has not been easy for me to wait, but I realize also that we must start with the best—with a man, who in the field of economics, means something of what Einstein and Veblen and Weyl mean in the field of mathematics. That we have now accomplished by being patient and being satisfied with nothing short of the best.

Bamberger responded with a warm handwritten acknowledgement, noting "the youthful enthusiasm that seemed to me to pervade the entire letter."[66] He made no mention of any additional financial support. Nor did Bamberger express any curiosity about Stewart's program. For his part, Stewart had been clear that his acceptance was conditioned on controlling supplementary appointments. Stewart informed Flexner that he expected Robert Warren to join him on the Institute faculty.[67]

Warren's resumé was quite thin for Institute consideration. He was born in 1891 in Plattsburg, New York. After completing undergraduate work at Hamilton College, Warren taught in Turkey for three years. Returning to the States he received a master's degree in economics from Harvard in 1916. Warren worked under Stewart for four years at the Federal Reserve Board, and then followed him to a Wall Street investment firm. With so much history as a subordinate of Stewart, Warren's credentials compared in some way to those of Mayer.

This time, Flexner quickly embraced an add-on appointment. As for the faculty already in the School of Economics and Politics, only Riefler knew anything about the negotiations. By Flexner's design, Earle and Mitrany were purposely kept in the dark. They would learn of their two new colleagues as *faits accompli*, following approval by the Board.[68]

Flexner was committing a significant breech of academic protocol. Mitrany and Earle were the senior professors in their school. Neither was given any voice in the selection of their three colleagues. These circumstances were unlikely to presage a harmonious environment. There were even serious doubts as to the prospects for collaboration between Riefler and Stewart. Following his return voyage with Stewart to the United States, Riefler wrote Flexner of his attempt "to work out a group which would be able to focus on all the varied problems of the economic scene from a rather unified point of view, mainly finance . . . . It is now clear, however, that I failed to convince Stewart." [69] Riefler, however, was a team player who went on to say "the most important thing we can do . . . is to make Stewart feel as happy as possible." Certainly Flexner was doing his part. Despite Warren's questionable credentials, both he and Stewart were offered the top faculty salary of $15,000.

Stewart and Warren accepted, to begin in January 1939. Clay decided to remain in England. The problem persisted of finding the money for two salaries. With Bamberger out of the picture, Flexner gave up hope of capitalization. He decided to appeal for a short-term solution from the Rockefeller Foundation, where Stewart was an insider. Would the foundation tide the Institute economics program over for a limited period? The response was encouraging, but noncommittal. When Flexner presented the nominations at the October 1938 Institute Board meeting, he took considerable liberty in overstating the financial arrangements: "I am, fortunately, in position to assure the Board that the requisite sum of money will be forthcoming whenever they enter upon their active duties."[70] It was another promise on which Flexner would be unable to deliver. With Stewart and Warren on the payroll for half of 1938-39, the Institute ran its first deficit, exceeding $25,000.[71]

There was no hyperbole in Flexner's comparison of his feelings over the acceptances of Einstein and Stewart. Whereas Einstein had launched the Institute, Stewart completed it. With the addition of Stewart, Flexner believed that the Economics faculty had reached the excellence of the other two schools. It was a view that, while sincerely held by Flexner, was open to dispute by more knowledgeable scholars.

The recruitment of Stewart paralleled another formative step for the Institute, the design of Fuld Hall. Flexner embarked on both projects in October, 1937, dependent on the same inadequate pool of resources and reluctant benefactor. Bamberger had already decided to sever his financial support. He intended to pay for a minimal building and then leave the Institute to operate on its endowment. To get the process started Bamberger donated $50,000.

During the initial planning, Mathematics expected to remain in Fine Hall and to set up a secondary operation at Fuld Hall. Even so, Veblen submitted a fairly extensive list of needs. When the requests for all of the schools were assembled, budget projections were clearly beyond what Bamberger had in mind. Flexner began to regard the building project as an albatross, competing for resources with the more important economics personnel search.[72]

Significant progress was made while Flexner was in Europe in the early summer of 1938. In separate meetings with Bamberger, Aydelotte and Maass attempted to close the funding gap. At first Bamberger appeared set on holding his additional contribution to $250,000. He was persuaded to add on another $100,000 while Maass was at work on the value engineering. When Flexner returned from Europe, the Fuld Hall plan was nearly worked out. The final steps were hampered by Veblen's insistence that faculty offices conform to the spacious 18 by 24 feet dimensions that he enjoyed in Fine Hall. Flexner, who was then scrambling to fund the new economics positions, had little patience for Veblen's nonnegotiable demands. Flexner took over as enforcer, and the plan was finalized in late September.[73]

The new building had long-term implications for the budget. Although Bamberger was handling the cost of construction, major infrastructure projects require funds for operation and maintenance. Since no money was added to endowment, Fuld Hall would be a further drain on income. The budget obligations were piling up.

During the 1935-1938 period Flexner was constantly torn by the conflicts between development, limited resources, and his own principles. Development won out. Five professors were added and construction of Fuld Hall was arranged. Along the way Flexner compromised many of his most cherished beliefs. Herzfeld was not full-time, the salaries of Goldman and Herzfeld were exploitative, Warren was unqualified, and the budget went into deficit. Flexner had maintained plasticity, but it was on his principles rather than design.

Fuld Hall cornerstone ceremony, May 22, 1939: from left to right, Alanson B. Houghton, C. Lavinia Bamberger, Albert Einstein, Anne Crawford Flexner, Abraham Flexner, John R. Hardin, Herbert Maass, and Harold W. Dodds. (Courtesy of the Archives of the Institute for Advanced Study.)

CHAPTER NINE

MOUNTING A COUP

With the arrival of Stewart and Warren in January 1939, the Institute faculty fulfilled Flexner's aspirations. The schools were balanced with robust rosters of five or six. The subjects reached each of the fundamental areas of the sciences, humanities, and the social sciences. Additional schools and professors were desirable, but none were essential. Moreover, no further development was feasible in the foreseeable future. For a weary 72-year-old director, who genuinely believed that his calling was to identify and recruit brilliant faculty, it was an ideal time for retirement.

Many on the faculty would have welcomed Flexner's departure. Veblen was frustrated by their increasingly acrimonious interactions. Some professors, such as Mitrany, Goldman, and Herzfeld, had been treated poorly. For most, Flexner's management approach was wearing thin.

Nine years as Director had dulled his sensitivities. In Flexner's original conception of the Institute, he proposed that the faculty have a substantial voice in decision making. Bamberger vetoed any statutory role, but informal involvement was not precluded. Early on, Flexner relied heavily on advice from Veblen, Frankfurter, and Morey. As the Institute developed, Flexner became more confident in his own wisdom, and more cynical over the benefit of faculty input. Among the sixteen professors, just a select few were ever consulted, and then only on certain issues. While this may have been an efficient approach to administration, it was problematic in certain areas.

The most contentious was in hiring, where Institute professors were better qualified than the Director to evaluate the merits of prospective colleagues. Earle, for example, was outraged that he was excluded from the selections of Stewart and Warren. It was not just the process that disturbed Earle. He viewed both Stewart and Warren as unsuitable for Institute professorships. Flexner did not understand the resonance of Earle's experience among the faculty.[1]

An inevitable area of faculty-director tension was the budget. Limited funds prevented Flexner from increasing salaries and supporting worthy initiatives. To the affected professors, Flexner often explained that their needs were a priority, but times were difficult. When prosperity returned, as he was certain it soon would, funding was assured. Yet some salaries were increased, new faculty were hired, and university projects such as Antioch continued to receive Institute money.[2]

Antioch fell under the heading of Institute-university relations, a sensitive area for Flexner. He had worked intimately with President Dodds and Eisenhart to forge a partnership that was crucial to the Institute's success. Flexner considered these dealings and their history to be a confidential matter that was outside the purview of the faculty. The policy was an outgrowth of Flexner's determination to avoid causing any embarrassment for the university. The greatest challenge to these efforts arose with regard to anti-Semitism.

Quotas on Jewish admission were prevalent then in American higher education. At Princeton University the Jewish presence was small. In the 1930s freshmen classes had enrollments of over 600 students, out of which the Jewish population typically numbered in the teens. Lefschetz was one of the first Jewish professors. The Institute brought a decidedly different demographic to campus. Six of the Institute faculty were Jewish.[3]

The Princeton community suddenly encountered a conspicuous Jewish component in its highest intellectual stratum. The anti-Semitic attitudes that emerged were disturbing to Institute professors. During the summer of 1934 Alexander wrote Veblen:

> When you get back I should like to take up with you the question of our relations with the University. I don't think they are anything like as rosy as Flexner thinks. People like Dodds have on their company manners when they talk to Flexner but usually let the cat out of the bag when they talk to me . . . . According to him, all the young people on the Princeton faculty are up in arms against the Institute, the German refugees, and the Jews.[4]

Flexner's reaction was to be expected. Regardless of what was conveyed by Dodds, Flexner was committed to portraying his partner institution in the best possible light. Moreover, as he himself was a Jew who had assimilated into refined traditional groups, Flexner was adept at denying discriminatory behavior. Several years later Earle failed in repeated attempts to raise similar issues about the Princeton community. Flexner viewed the anti-Semitism card as the third rail for the relations he was cultivating.[5]

Altogether the issues of hiring, budget, university relations, and anti-Semitism contributed to a significant reservoir of faculty discontent. The Stewart and Warren appointments triggered a call by some professors for a faculty role in decision making.[6] Flexner reacted by raising the governance procedures in his January 1939 report to the Board. Despite his recent covert operation in the School of Economics and Politics, Flexner began by stipulating that each of the three schools was autonomous and self-governing. He went on to say that there were limits, however, to the usefulness of faculty involvement.

> It would be a waste of time and a hindrance to effective progress if the three groups were regarded as a faculty which met at stated intervals and legislated for the entire Institute . . . . at long intervals some point of general interest may arise on which the faculty should be brought together and consulted, and its views or conflicting views should be transmitted to the trustees, but anything more than this would be a waste of time and energy and would be the first step in forming a routine which might ultimately choke what is today the outstanding merit of the Institute.[7]

As for the role of the Director:

> It may at any time be his most important function to have the final word—after conference inside and outside the Institute—in the matter of faculty appointments, though the presumption is strong that the members of a given school are the best and the proper judges. It is also the business of the Director, using such income as is annually available, to enable the scholars, who really are the Institute, to do, in so far as is humanly possible and reasonable, what they themselves regard as important. That is, of course, a more complex and delicate task when there are three schools . . . .

He concluded with the familiar refrain that under the unique organization of the Institute

> we have lived happily and cooperated effectively so that professors, members, and secretaries have from the beginning formed a happy and efficient group, every member of which has been interested not only in his or her work, but in the welfare of the Institute and the cultivation of friendly and helpful relations with Princeton University . . . .[8]

The grievances of the faculty were obscured by Flexner's framing of the issue. Discussion by the trustees focused on the question of whether faculty meetings were desirable. No real consensus emerged. Flexner concluded the conversation by saying grudgingly that he "had no objection whatsoever to the faculty's meeting whenever it pleased but" regularly scheduled meetings were, in his judgment, counterproductive. If the faculty did choose to meet, then he would not attend unless the meeting were called for some particular purpose requiring his presence.

In contrast to the extensive coverage of the discussion on faculty assemblages, the January 1939 Board minutes carried no mention of a significant decision taken around this time. The School of Mathematics was to relocate in Fuld Hall. The record is silent as to what caused the Institute to abandon Fine Hall and where the initiative originated. Given Flexner's sensitivity to relations with the university, he may have concealed the topic from the trustees, or suppressed it from the minutes. In any event, shortly after the Board meeting, arrangements were being made for Mathematics to occupy Fuld Hall when it opened in the fall.[9]

Reconciling the space needs of the three schools was difficult. To address these issues Flexner called a meeting of the Institute faculty. Among the topics were room allocation and preserving connections to the university. In the midst of the discussion, Alexander raised his allegations of anti-Semitism at the University. Flexner was blindsided. He had deluded himself into believing that everyone, except Mitrany, was a happy member of the community. For Flexner the meeting was far worse than a waste of time. It was harmful. He would never again call together the faculty.[10]

The early February meeting provided the impetus for members of the faculty to organize in a crude way. One week later they gathered on their own at an informal dinner. Professors shared their views and experiences on the operation of the Institute. Frustration was widespread.

Nor was Flexner having a good time. Rather than searching for another genius, he was continually quibbling with Veblen over the assignment of each square foot in Fuld Hall. Flexner's other nemesis was the budget. For the economics program he had counted on multiyear support from the Rockefeller Foundation, but personnel changes there had resulted in a moratorium on new social science grants. Flexner needed about $25,000 to avoid a deficit in the current year. Bamberger pointed Flexner to the Rockefeller Foundation, which sent him back to Bamberger.[11]

Flexner understood the workings of the Rockefeller Foundation. Matching funds held a special appeal, and short-term grants could be made administratively. Out of desperation, Flexner concocted a disingenuous scheme to achieve balance over the remainder of the year. Bamberger agreed to offer $25,000 under two stipulations. An equal amount was to be provided by the

Rockefeller Foundation, and any unused portion (likely $25,000) was to be returned to Bamberger. There was little chance for funding if the Rockefeller people understood that Bamberger was merely flashing his money. So when Flexner presented the bizarre proposal to the Foundation, he made it appear. as if the donor was some unnamed person other than Bamberger. The ploy failed. An administrative allocation was declined.[12]

Flexner was left with no alternative other than to swallow the $25,000 deficit, and wait until the new Rockefeller program director was ready to evaluate proposals. With only Fuld Hall on his plate, Flexner decided to take a trip to the Bahamas for a rare winter vacation. Throughout his tenure as Director, Flexner had spoken of retirement as a medium term event. It was always a few years in the future, but never assigned a date. Correspondence before and after the vacation indicate that Flexner was giving serious consideration to stepping down.

He still had Aydelotte in mind as his successor. Flexner asked Aydelotte to draft some remarks about the history of their friendship and interactions. Prior to his departure Flexner received a long, laudatory reply. When he returned from the Bahamas, there was a letter from trustee Alanson Houghton expressing his desire to retire as chairman of the Institute Board. Flexner replied that he also was contemplating retirement. No date was given, making the likely time frame on the order of a year or two.[13]

Whether or not Flexner had given faculty members some indication of the impending event, there was a reasonable inference that the time was approaching. The hypothetical retirement was among the topics discussed by the faculty at a second gathering. The choice of a new Director would affect the lives and work of everyone. Some professors were anxious to assert that the faculty hold a consultative role in the selection of Flexner's successor. Warren argued that the faculty as a body lacked the standing to participate in Institute decision making. The professors discussed these and other issues over their dinner in mid-March. Of some guidance were the minutes of the January trustee meeting that had been disseminated to them. In his Director's Report Flexner had noted that there were circumstances under which faculty views should be transmitted to the trustees.[14]

Einstein, Goldman, and Morse were appointed to communicate the faculty's position to Flexner. Einstein was chosen for his clout, rather than letter writing skills. Since their late 1933 reconciliation, Einstein had managed to get along with Flexner. The absence of further confrontation did not mean that Einstein was content with Flexner's leadership. Tolerance of anti-

Semitism was among the sore points.[15] That Einstein was willing to accept the role as faculty representative confirms that he was among the more activist element. The Committee's letter to Flexner is reproduced in full below:

> You have been kind enough to send the faculty your report made to the Trustees at a meeting on January 25, together with the comments which the members of the Board made thereon. At a recent informal dinner certain aspects of this report were discussed by the professors of the Institute, and we were requested to give you an account of the conclusions reached.
>
> The Institute has now developed in its three schools to a point where its character can be clearly seen and appreciated, and the most important problem from now, in our eyes, is the stability of what has been achieved by the generosity of the donors and your own creative insight.
>
> This stability will depend upon the wisdom and deliberation with which future Directors are chosen. It is the unanimous opinion that this choice should be preceded by a preliminary consultation with the faculty.
>
> It is equally essential in the opinion of the majority of the faculty that no professor be appointed without a similar consultation with his future colleagues.
>
> We understand that both the responsibility and the final choice in each case rest with the Director and the Board of Trustees. Their action should, however, in our opinion, be preceded by a consultation with the faculty which should be made effective by allowing adequate time for the consideration and inquiries which are necessary in each case.
>
> The professors earnestly desire that the above conclusions be conveyed to the Board of Trustees. We should like very much to talk these matters over with you, and to add any information which you may desire concerning the opinions expressed.[16]

The faculty were proceeding cautiously. Their respectful letter followed the guidelines laid out by Flexner. The requests were reasonable and consistent with academic tradition. It is notable that unanimity was reached on the director but not the faculty issue.

Flexner reacted defensively to the faculty requests. He immediately drafted a response to Morse. In this letter Flexner rehearsed his tireless efforts to obtain advice at every stage of Institute development. He lamented that any codification of hiring procedures would needlessly inhibit the flexibility that he had exercised so successfully. "The members of the faculty are the natural and logical advisors and consultants to the director and the trustees, but they have no vested rights, and, in my judgment, they would be very unwise to ask for a vested right or privilege."[17] Flexner added that the faculty's activism was likely to disturb the founders. The letter was not sent. Instead, Flexner summoned Morse and Einstein for individual interviews.

Both Morse and Einstein were frustrated by their meetings with the Director. Flexner was unresponsive to the bottom line question of whether he would forward the faculty's letter to the trustees. It is clear that he had no intention of doing so. Flexner's attitude was that if he did not indulge the faculty's bad behavior, they would come to their senses and the phase would pass.[18]

Flexner had every reason to believe that his paternalistic approach would succeed. What could the faculty do? Authority was vested with the Director and the Board. The faculty's access to the Board was through the Director. Moreover, the trustees had never been inclined to cede power to the faculty.

The professors had been maneuvered into a bad position. The only route around Flexner's stonewalling was a clandestine approach to the trustees. It was a course fraught with danger. Members of the Board might refuse even to hear a direct appeal, let alone decide in its favor. Flexner was almost certain to learn of the faculty's action and regard it as mutinous. Indeed, once undertaken, the objective had to be the overthrow of the Director. If the coup failed, Flexner could be expected to make life difficult for the conspirators.

No possibility existed of a unified faculty offensive. Meritt, Riefler, Stewart, and Warren supported the Director whose reign had become intolerable to others. Each professor faced decisions on where he or she stood and whether to act. Records reveal that Einstein, Veblen, Earle, Goldman, and Mitrany took their grievances to individual trustees. Other faculty members may have followed the same course and succeeded in covering their tracks.

The risk was not evenly distributed among those who challenged Flexner. Einstein and Veblen were bulletproof. Their careers were complete and their $8,000 pensions guaranteed. Goldman and Mitrany were already at odds with Flexner and they probably saw little more that he could do to them. Earle's situation was radically different from his coconspirators.

Earle was vulnerable. He was in the second year of rebuilding a career that many thought was over. Just a few months earlier a throat infection had forced him to cancel a research trip to California. Earle had already lost too much time. He could ill afford to divert any energy from his work. Nor could his reputation, still clouded by health issues, stand up to allegations of troublemaking.[19]

Even more compelling than Earle's rational considerations were the emotional ones. His debt to Flexner could never be repaid. Flexner had always been there for him with support in every possible way. The constant letters

and family help were uplifting. The Institute job offer, under the circumstances, was a remarkable act of human devotion and faith. Flexner was the
last person on earth whom Earle should want to betray.

The friendship between Flexner and Earle went back to the mid-1920s.
Early on Flexner had shared with him the concept of the Institute. Earle was a
good audience. A devoted scholar who held an idealistic view of the academy,
he embraced Flexner's proposals for elevating research and education. In his
sickbed Earle was thrilled by Flexner's updates on the development. Year after
year Earle anticipated overcoming his disease and joining the paradise for
scholars.

It probably would have been impossible for the Institute to live up to these
expectations. For Earle it did not come close. How could an utopian intellectual community deny the obvious presence of anti-Semitism in its midst
while having its research direction driven by uninformed views of economics?
Earle tried to raise these issues with his old friend whom he owed so much.
They spoke often, but Earle was unable to reach him. In these conversations
Flexner constantly denigrated Mitrany. Earle cringed at each unprofessional
slur of a scholar whom he respected. Then came the secret hiring of Stewart
and Warren, an act antithetical to Earle's expectations of an academic community.

When Earle discussed his experiences and perceptions with Institute colleagues, it became clear that the problems transcended the School of Economics and Politics. Earle informally joined other professors in seeking a
faculty voice in Institute decision making. When Flexner pocket vetoed the
letter to the trustees, it was the last straw. Earle became part of the revolution,
never regretting the betrayal of his champion.

Ousting Flexner would not be easy. Each year at its spring meeting, the
trustees voted on the Director's reappointment for the coming academic year.
Normally, the spring meeting was held in April. In 1939 it was pushed back
to May 22, in hopes of obtaining pleasant weather for a Fuld Hall cornerstone
ceremony. This left two months, after the Morse and Einstein meetings, for
the faculty to lobby the trustees.

The fifteen-member Board included Veblen, Riefler, and Stewart, whose
positions on Flexner were known. To obtain a majority, the faculty needed to
persuade seven trustees to oppose reappointment. Veblen understood, however, that the Board did not function democratically. Controversial matters
were decided by Bamberger. The trustees with access to Bamberger were
Hardin, Maass, Leidesdorf, and Flexner. The targets of the faculty campaign

would be Maass and Leidesdorf. Without their support, there was no chance of forcing Flexner out of office. The faculty strategy also included two other influential trustees, Frank Aydelotte and Lewis Weed. As academic administrators themselves, Aydelotte and Weed were likely to be more attuned to the issues being raised by the faculty. If they could be convinced that Flexner needed to go, the faculty would have two important allies in winning over Maass and Leidesdorf.

The normal channel for faculty-trustee communication was through the Director. If a professor complained to a trustee, the proper response for the trustee was to urge the professor to work out the problem with the Director. The faculty faced the likelihood of their grievances being dismissed out of hand. It was decided that approaches were to be restricted to trustees with which the professor had a pre-existing relationship. For example, it happened that, at Aydelotte's invitation, Einstein had delivered the Swarthmore commencement address the previous year. On March 26 Einstein wrote to Aydelotte. Enclosed was a copy of the faculty's letter to Flexner, accompanied by a cover note in which Einstein described his unsatisfactory conversation with Flexner. The following day Mitrany wrote Aydelotte, invoking Einstein and requesting an interview.[20]

Aydelotte was in an awkward position. As a scrupulous adherent to academic protocol, he would not normally allow himself to be placed between a professor and the Director. Recall that Aydelotte had upbraided Frankfurter for communicating directly with Riefler. Yet Einstein was not an ordinary professor, and his Swarthmore visit had meant a great deal to the college and its president. Aydelotte permitted Mitrany to brief him on the faculty's dissatisfaction with Flexner. It quickly became clear that the dissident faculty could only be placated by Flexner's retirement.[21]

These circumstances made Aydelotte's involvement even more problematic. Not only did Aydelotte know that he was Flexner's choice to be the succeeding Institute director, but Aydelotte also wanted the job. Associating himself with a movement to depose Flexner, regardless of its merits, had the strong appearance of opportunism and impropriety. Aydelotte decided to contact Weed at Johns Hopkins, forwarding the information from Einstein.[22]

Weed had already been visited by his fellow trustee Veblen. Veblen presented the case against Flexner and urged Weed to speak with Earle, who was coming to Hopkins for a medical examination. Weed did not connect with Earle, but they did communicate through a medical school physician. Both Veblen and Earle were effective advocates who were preaching to a receptive audience. For years Weed had suffered through Flexner's self-aggrandizing presentations and domineering tactics with the Board.

Weed replied to Aydelotte on April 11:

> I think we are facing a very serious situation. The matter is one of funda-
> mental academic organisation as well as the involved question of retirement
> of the Director and appointment of a successor. We must get together soon
> so that we can have a long and frank discussion of the matter before our
> next meeting.[23]

Both Weed and Aydelotte were listening to the faculty complaints.

Aydelotte did meet with Weed and then with Maass and Leidesdorf. Einstein had made the initial approach to Leidesdorf, virtually guaranteeing the
message would reach Maass. Einstein's effect on these two trustees cannot
be overestimated. When Maass and Leidesdorf discussed the Institute with
their friends, names such as Weyl and Panofsky made no impression. It was
Einstein that everyone wanted to hear about. Einstein was the one Institute
professor whose discontent would register with Maass and Leidesdorf.[24]

It is impossible to reconstruct all the moves that were occurring behind
the scenes, but Maass and Leidesdorf were coming around. On May 4 Veblen
wrote Aydelotte:

> I saw Maass and Leidesdorf however last night, and talked things over with
> them quite fully, and from my point of view satisfactorily. I got the impres-
> sion that the discussions which have now taken place are probably sufficient
> for the present.[25]

This indication of Maass' and Leidesdorf's alignment with the dissident
faculty was a crucial development. No doubt Einstein's clout and Veblen's
persuasive skills were powerful forces. Yet Maass and Leidesdorf were cautious professionals who were suddenly brooking a radical movement for the
overthrow of a long-time associate. This departure from traditional behavior
calls for a closer examination of the two trustees' relationship with Flexner.

The introduction of Flexner to Maass and Leidesdorf was the historic
consultation out of which the Institute was conceived. The follow-up actions
by Maass and Leidesdorf brought together Flexner and the founders. During
the Institute's formative years Maass and Leidesdorf served substantive intermediary roles as confidantes and advisors to both principles. Leidesdorf was
the Board's first treasurer. When the founders retired as trustees, Flexner had
Maass installed as vice chairman. Through this position, his own initiative,

and his close relationships with Flexner and Bamberger, Maass became the most powerful member of the Board.

As was common among friends of their standing, Flexner, Maass, and Leidesdorf often exerted their influence to advance the careers of one or another's children. Maass used his Congressional connections to assist Flexner's son-in-law obtain a Federal job in Washington. In mid-1937, both Leidesdorf's and Maass's sons were applying to college. Flexner pulled many strings on their behalf. Despite these efforts, the trustees' sons were turned down at the University of Pennsylvania and Princeton respectively. With the assistance of Flexner and Meritt, Maass' son eventually gained admission to Brown.[26]

Maass was unable to reconcile himself to the Princeton rejection. A few months later he and Leidesdorf expressed concern to Flexner about the presence of anti-Semitism in the Princeton community. They feared that, in the future, the Institute might be susceptible to its spread. Flexner was horrified that such accusations were coming from the Board. It is easy to imagine him frantically attempting to lower the volume of the conversation so that it would not be overheard.[27]

Flexner did his best to disabuse Maass and Leidesdorf of their concerns. As always, Flexner substantiated his position by citing his own vast knowledge of American higher education. He explained that enlightenment and discrimination were incompatible. The arguments did not persuade Maass and Leidesdorf. They had their own experience with anti-Semitism in New York and with admission quotas in universities. Flexner had little credibility with them on this issue. In their very first meeting, Flexner had denied the existence of discrimination against Jews in medical schools. Over the years Maass and Leidesdorf had come to know Flexner well and observe his attitude toward anti-Semitism.

The conversation continued during the 1937 Christmas period. A trace of Veblen's fingerprints is suggested in some of the measures Maass and Leidesdorf advocated to protect the Institute against anti-Semitism. The proposals included the election, by the faculty, of representatives to the trustees. Another suggestion was for the School of Mathematics to evacuate Fine Hall for Fuld Hall (this was a year before the decision was made). Flexner strongly objected to everything he heard. After three meetings, no minds were changed. Flexner was terrified that if Maass and Leidesdorf's views got out, the consequences for the Institute would be catastrophic.

Flexner decided to act pre-emptively to inoculate the other trustees from the dangerous views of Maass and Leidesdorf. First he consulted with Ay-

delotte and Weed. Aydelotte expressed confidence that the Institute Board
and faculty were in no danger of anti-Semitic infiltration. Unlike Flexner,
Aydelotte did acknowledge the presence of anti-Semitism at Princeton, and
advocated, for entirely different reasons, an increased faculty influence on the
Board.[28]

Neither Weed nor Aydelotte were present at the January 1938 Board
meeting, which was attended by both Maass and Leidesdorf. Toward the end
of his report to the trustees, Flexner abruptly digressed to the topic of discrimination. Without specifically mentioning Judaism or anti-Semitism, he
contrasted the extent of racial and religious prejudice in the world, America,
and the academy.

> The atmosphere in which the world lives has distinctly deteriorated since
> the first steps were taken looking to the creation of the Institute . . . .
> There is no fitter arena in which this battle of decency and tolerance can be
> fought out than is furnished by the institutions of learning. I have myself
> no fear for the future of American universities on this score. Faculties—and
> I speak from a wide and intimate acquaintance with them in all sections
> of the country—have practically without exception long since risen above
> denominational or racial prejudice, but as a matter of fact decisions unfa-
> vorable to this or that person are often based upon merely the enforcement
> of high standards, and it is frequently a face-saving gesture on the part of
> the unfortunate individual to attribute his ill success to intolerance.[29]

Some of the trustees may have been heartened to receive Flexner's absurd
verdict that, in 1938, American universities were essentially free of religious
and racial discrimination. For Maass and Leidesdorf, who knew better, even
more disturbing was the unmistakable reference to themselves and their sons.
It was humiliating to sit and listen to their motives being impugned and their
children ridiculed, albeit anonymously. Maass and Leidesdorf were the victims of an insensitive act by an ally who had previously treated them with
complete solicitude.

It was with the background of these experiences that Maass and Leidesdorf heard the faculty complaints about Flexner. The faculty argued that the
72-year-old Director had lost his grip and was no longer able to handle the
demands of running the Institute. Both this argument and the specter of
anti-Semitism were concerns that could be persuasive with Maass and Leidesdorf. For his part, Flexner was oblivious to the whispers among the faculty
and trustees. He saw no immediate reason for retirement. His health was
good and the funding problems in economics required his attention. When

the slate of nominations was arranged for the May, 1939 meeting, Flexner expected his name to go forward for another year as Director. As for any opposition, knowledgeable trustees operated under Bamberger's rules. Institute Board meetings were not battlefields. Maass, Leidesdorf, and Veblen would have to take on Flexner in a more private manner.[30]

At the May Board meeting Flexner was reappointed for an additional year. The Director began his report by briefly commenting on the faculty response to the discussion of the previous meeting. The letter from Einstein, Goldman, and Morse was not mentioned. Flexner merely reported the faculty as in agreement with his own expressed views, adding: "Nothing in the internal situation of the Institute or in world conditions now calls, in my opinion, for any change whatsoever. The Institute is a happy group of scholars."[31] It must have been difficult for Maass, Aydelotte, Weed, and Veblen, who were all in attendance, to mask their incredulity.

The trustees were presented a 1939-40 budget with a deficit of $37,000. The lines for member stipends were severely cut back. Taken together with the deficit that was by then assured for the current year, the Institute faced serious financial problems. The Board approved the budget without any discussion appearing in the minutes. After the meeting was adjourned, the trustees traveled to the Fuld Hall construction site for the cornerstone ceremony. It was a low key event. Representing the founders was their sister Lavinia Bamberger. The principal speaker was President Harold Dodds of Princeton University.[32]

Dodds' customary platitude about the establishment of the Institute's physical structure was in contrast to the shaky financial foundation on which it all rested. The existing fiscal situation was unsustainable. The Institute's repeated deficits were a clear manifestation of inadequate endowment, and there was no expectation for an improvement in the foreseeable future. With the sorry state of the economy over the past decade, it was unrealistic to count on a better return from investment. Moreover, other trend lines were ominous. The current reduction in the number of members was an emergency excision of bone rather than fat. Payouts for unfunded pensions were on the horizon.

Flexner was still pursuing a Rockefeller Foundation grant for the program in economics. Although he had expected the money to be in place several months earlier, Flexner remained optimistic. He was convinced that Stewart's and his own insider connections would succeed. It should be noted, however, that even if the funds were approved, the Institute's shortfall was merely deferred. When the grant expired, the problems of an inadequately capitalized institution would resurface.

Maass and Leidesdorf were the trustees who most closely tracked the Institute's finances. As treasurer, Leidesdorf held oversight responsibility and frequently discussed the budget with Flexner. Maass had recently attempted to address the pension problems, only to be rebuffed by Flexner. Through these experiences Maass and Leidesdorf observed Flexner's increasingly ineffectual handling of the budget. As the financial difficulties were mounting, the anti-Semitism schism developed and the faculty discontent emerged.[33]

The question facing Maass and Leidesdorf was whether or not to take the drastic step of attempting to remove the Director. If they did not act, Flexner would remain Director for at least one more year. Either course held the possibility of embarrassing publicity for the Institute. In the early summer of 1939 Maass and Leidesdorf conducted two interviews with pairs of dissident faculty. One meeting was with Einstein and Earle, the other with Goldman and Veblen. All four professors were fed up with Flexner and pulled no punches. They were adamant that Flexner had to go.[34]

Maass and Leidesdorf agreed. The cleanest scenario was for Flexner himself to realize that the time had come to step down. On June 5 Maass and Leidesdorf attempted an intervention. They met with Flexner and urged him to retire. The move took Flexner by surprise. He was incapable of recognizing the seriousness of the Institute's problems and the extent of the faculty opposition. Convinced that Maass and Leidesdorf were overreacting to the unreasonable complaints of a malcontent or two, Flexner refused to provide an immediate response.[35]

Just as eleven years earlier, when associates at the Rockefeller Foundation had pressured him to retire, Flexner believed that he was performing his duties in an exceptional manner. To confirm that the call for retirement was unwarranted, Abe sought advice from his brother Simon and what he later characterized as "several disinterested persons here in Princeton."[36] The outcome was predictable. Simon had long realized that he could not administer *tough love* to his younger brother. Abe could only be left to choose his battles and live with the consequences.[37]

As for the other consultants, they were not exactly disinterested. Flexner quickly sought out Meritt and Earle for advice. The professors, both of whom held filial ties to Flexner, were requested to offer frank assessments of any existing problems at the Institute. The Earle interview was arranged for a luncheon that included Mrs. Flexner, hardly a setting for serious criticism.[38]

Earle was conflicted over whether to honor his acceptance of the lunch invitation. He had previously attempted, at his own initiative, to communicate the requested information to Flexner, but never succeeded in piercing the Director's shields. Now, having been forthright with the trustees, it would be hypocritical to play any further role in Flexner's self-deception. Yet Earle was not up to the bluntness required for his criticisms to get through to Flexner in a face-to-face meeting. Earle sought advice from Maass, who suggested that he decline.[39] At the last minute Earle sent his regrets, along with a letter unmistakably conveying his views. The main paragraph enumerated most of the faculty grievances:

> You asked me yesterday that I tell you the truth without fear or favor. As a matter of fact, that is precisely what I have been trying to do in innumerable conversations during the past three years. I have expressed to you my alarm on a number of points, more specifically: your policies vis a vis Princeton University; your refusal to admit the existence of anti-Semitism in this community; your openly expressed contempt for fellow-members of the Faculty, sometimes taking the form of personal abuse; your insistence upon dealing with us (except the Mathematicians) always as individuals not as members of several schools or of the Faculty as a whole; your resistance to a measure of Faculty participation in vital decisions; your refusal to transmit to the Trustees a respectful and modest request for such participation; your procedure in the most recent appointments in the School of Economics and Politics, which violates every tenet of long-established and universally respected principles of scholarly communities; your marked favoritism toward individuals (including, doubtless, me myself) and toward certain subjects, notably economics; an increasing tendency to make ex parte decisions.[40]

Flexner was furious, both at being stood up and at the impertinence of the judgments. After all that he had done for Earle, Flexner expected more consideration on both counts. The substance of Earle's criticisms were never taken to heart. Flexner framed them for more sympathetic consultants, such as Stewart and Dodds, to rebut.[41]

Meanwhile, Maass and Leidesdorf were awaiting a response. Flexner was in no hurry. His delay was driven by a combination of passive-aggressiveness and a genuine struggle to reconcile considerations pushing him in different directions. The notion of retirement was appealing. Relinquishing the duties of Director would finally afford Flexner some leisure and the opportunity to draft his memoirs and a long-planned book tribute to Gilman. Flexner weighed these benefits against the humiliation of being coerced, for a second time, into retirement. He sought an exit strategy that could not be construed, by anyone, as capitulation to his enemies.[42]

It was easy for Flexner to satisfy himself that retirement was of his own initiative. After all, he had recorded his intention in the letter to Houghton that predated the meeting with Maass and Leidesdorf. However, a sudden departure lacked dignity and might give satisfaction to Maass, Leidesdorf, and Earle. To avoid any doubts as to whether he was leaving on his own terms, Flexner was inclined to serve out the coming academic year, for which he had already been appointed by the Board. A retirement announced one year in advance permitted an orderly transition while heading off the type of questions raised by his precipitous withdrawal from the Rockefeller Foundation.

As Flexner pondered his future, he maintained silence with Leidesdorf and Maass. When two and one half weeks had passed since their meeting, Leidesdorf and Maass decided the time had come to press for an answer. First they sought clearance from Bamberger. Maass then phoned the stonewalling Flexner.[43]

Later that afternoon, Flexner pitched his deferred retirement proposal to Veblen. Flexner began the conversation by broaching the letter from Earle. Veblen acknowledged some familiarity with Earle's actions, but refused to be drawn into a debate over specific grievances. Veblen also disclosed that he was aware of, and endorsed, the advice offered by Leidesdorf and Maass. Veblen recommended to Flexner that he retire immediately. Whether or not Flexner had realistically expected a different outcome, the discussion did confirm for him that he faced a conspiracy consisting of Veblen, Earle, Leidesdorf, and Maass.[44]

In the weeks and years that followed, Flexner insisted that the opposition was no more widespread than these four. There was something in Flexner's psychology that made it important to place the bound on the number of faculty and trustees seeking his removal. Even Mitrany was omitted from the rants. While Flexner denied the extent of the opposition, he did recognize that an attack was underway.[45]

From his conflicts on the General Education Board and at the Institute, Flexner was no novice at power struggles. He moved to shore up his support by contacting John Hardin, the only seemingly uncommitted trustee remaining from Bamberger's inner circle. There was a long-standing enmity between Hardin and Maass with which Flexner was well acquainted from dealings over the past decade. It was easy to manipulate Hardin into deprecating any action of Maass. Flexner left the meeting confident that Hardin was firmly on his side.[46]

Flexner was ready to play hardball to remain on as Director. He would not demean himself by negotiating with his enemies. Why should he concede to Leidesdorf and Maass that they held the standing to rule on a retirement date? Flexner was especially incensed at Maass and Veblen, whom he regarded as the instigators. A reply to Maass was overdue. Yet Flexner could not bear the thought of dictating the salutation. On June 26 Flexner sent the following curt note to Leidesdorf, copying Maass:

Dear Mr. Leidesdorf:

Since my last conversation with you and Mr. Maass in your office on June 5 I have not only given very thoughtful consideration to what was said on that occasion but I have conferred with my brother Simon, with several disinterested persons here in Princeton, and last Friday I laid the whole matter before Mr. John R. Hardin, asking for his counsel. Mr. Hardin's advice coincided with that which I had previously received from my brother and others. It was to the effect that at the present time I should take no step whatsoever.[47]

Flexner did not even record his intention to retire. This tactic left him plenty of flexibility for future maneuvers. Time was also on his side. Posting the letter was one of Flexner's last acts prior to joining his family in Ontario for the remainder of the summer. In the Canadian woods he was well positioned to evade further communication, stalling his betrayers until the fall. The effort for an amicable change in Institute leadership had failed. Forcible removal would be difficult.

Under these circumstances Leidesdorf and Maass might have been wise to settle for the deferred retirement arrangement that Flexner had presented to Veblen, assuming it were still available. At least this would establish a date certain for Flexner's departure. The real losers would be the opposition faculty, who could expect to endure a year of retribution. The qualms of these victims were largely muted. They had evacuated Princeton to escape the oppressive seasonal humidity. Veblen was at his summer home in Maine, Einstein on Long Island, Mitrany in England, and Earle taking in the healthy air from the Adirondack region of New York State. Communication barriers obstructed the professors from uniting to press their case with Leidesdorf and Maass.

That is not to say that no one was lobbying Leidesdorf and Maass. Stewart dismissed his colleagues' complaints as "a tempest in a teapot." The primary voice of the disgruntled faculty was Earle who had been liberated by his vent to Flexner. After the cancelled luncheon Earle sent a steady stream of messages to Maass and Aydelotte, seeking news and urging vigilance in the effort to depose Flexner. By stepping forward at this time, Earle became the de facto spokesperson of the dissident faculty to the trustees. Of particular significance were the exchanges between Earle and Aydelotte.[48]

Aydelotte was in the influential position of being a respected confidant of key players from each constituency, including Earle, Maass, Flexner, and Bamberger. By this time Aydelotte had reconciled himself to the necessity of removing Flexner from the Directorship. On June 24 Aydelotte discussed developments with Earle over the phone. It was the day after the Flexner-Hardin meeting. With Aydelotte's connections, he knew that Flexner had decided to put up a fight. Aydelotte conveyed the bad news to Earle. They then strategized over how, most effectively, to portray the faculty discord to Leidesdorf and Maass. Aydelotte asked Earle to predict the outcome of a faculty vote of confidence in Flexner. The principled Earle was uncomfortable with taking on such a grave responsibility. He asked for additional time to think it over, responding the next day in a letter with the following earnest answer:

> Assuming, however, that a committee of the Trustees were to call the Faculty together and discuss the situation as frankly as Maass and Leidesdorf discussed it with me, I think the following would without hesitation vote yes on a resignation: Einstein, Veblen, Morse, Alexander, von Neumann (personally indifferent but loyal to the group), Weyl, Mitrany, Goldman, Herzfeld, Earle, and probably Panofsky. Unreliable under pressure, but thoroughly fed up with the prevailing state of affairs: Lowe (whose proxy Einstein holds). This leaves Meritt, Riefler, Stewart, and Warren, who are, as you know, in a special category.[49]

Earle went further. Earlier in the month he had driven Einstein to New York for their meeting with Leidesdorf and Maass. While in transit the two professors discussed the crisis at the Institute. Einstein raised the possibility of going elsewhere if Flexner were to remain on as Director. It is difficult, today, to ascertain Einstein's true intentions at the time. He may have been indulging Earle's outrage, merely letting off steam, or, possibly, he was serious. In any event, Earle was torn between his desperate desire to impeach Flexner and the propriety of interpreting, for others, a remark from a personal conversation. In his draft to Aydelotte, Earle typed: "I know, for example, that Einstein has about decided to resign in October unless there is a new Director."[50] Before mailing the letter to Aydelotte, Earle crossed out the word "know" and wrote in "think."[51]

Earle found himself confronting other ethical considerations. Aydelotte asked for a copy of the June 9 bill of particulars that Earle had sent to Flexner. Although Earle had labeled the note "strictly confidential," he reluctantly decided to make copies for Aydelotte and Maass. Earle's guilt was assuaged when he later learned that Flexner had already divulged the secret contents.[52]

Accompanying Earle's principles of right and wrong was a degree of naivité. He assumed that the trustees were receptive to giving the faculty a greater role in Institute governance. Earle constantly made suggestions to Maass and Aydelotte about faculty participation in the selection of a new Director. Some of these conversations were especially awkward for Aydelotte. Earle believed that Flexner was plotting with Stewart for a handover of the directorship. As Aydelotte knew, he, rather than Stewart, was Flexner's designated successor.[53]

In late June Aydelotte discussed the Directorship situation with Leidesdorf and Maass. The three trustees were in agreement that the welfare of the Institute necessitated Flexner's removal. Yet it was clear that Flexner would not go voluntarily, and no good option presented itself. Forcing him out required an unseemly action by the Board that might be open to legal challenge. Flexner would certainly be responsive to a request from Bamberger, but the 84-year-old founder was insulating himself from the coup. Bamberger had lost confidence in Flexner's judgment and favored his retirement. However, Bamberger held a sentimental attachment to his loyal Director, and insisted that the message be delivered by others.[54]

A promising channel opened when Flexner invited Aydelotte to Canada for a visit in early July. Leidesdorf and Maass urged Aydelotte to make the trip. Perhaps he could succeed where they had failed. The task was monumental. Earle, Leidesdorf, and Maass had experienced the difficulty of reaching Flexner with a critical message he did not want to believe. Getting through required such tactless candor that Flexner's attention abruptly shifted to despising the messenger. As has been said of Thomas Jefferson, Flexner was so adept at self-deception that he "could have passed a lie-detector test."[55]

Aydelotte had good reason to decline. Time was a scarce commodity. The busy college president was already trying to fit trips to Mexico and Europe into what remained of the summer. Moreover, in seeing Flexner, Aydelotte was being called on to play a disingenuous role. He was to arrive under the guise of being a supporter, accept the hospitality of an old friend who was under siege, and then work to achieve the objective of Flexner's enemies. It was a distasteful mission, but it appeared that Aydelotte was the only hope for helping Flexner to realize what was in his best interests.

Aydelotte and his wife traveled to Canada, where they were welcomed by the Flexners. Abe loved the "camp." There were magnificent views of a lake and woods. Daily activities included fishing, swimming, sawing, and rowing. Away from the pressures of office life, he relaxed with his family and guests. It seemed an ideal environment for a respected friend to employ his powers of persuasion. Aydelotte also had a valuable ally in Flexner's wife, Anne. She had reached her own conclusion that it was time for her husband to retire. Aydelotte and Anne did their best, but it was to no avail. Flexner seemed oblivious to their urging. He was determined to remain on as Director for another year, and then pass the torch to Aydelotte.[56]

Returning home, Aydelotte briefed Earle on Flexner's intentions, omitting the plans for a successor. Aydelotte seemed resigned to Flexner having his way. Earle was discouraged. Six weeks had accomplished nothing. If Aydelotte was caving in, there seemed little hope that Leidesdorf and Maass would persevere on their own. Earle needed help. He appealed to Veblen, who, since leaving Princeton, had remained out of the loop. Earle also wrote to Aydelotte, warning of dire consequences if Flexner were left to his own devices. Upon hearing from Earle, Veblen re-entered the picture, endorsing the position of Earle.[57]

With some distance from Flexner, Aydelotte regained his perspective. After meeting with Maass, Aydelotte reported their agreement that Flexner needed to step down. What troubled Earle was that neither trustee seemed resolved to taking stern measures to oust Flexner. The current plan was for Hardin to speak with Flexner. The summer was slipping away and Flexner's stalling tactics appeared to be succeeding. Earle was terrified at the prospect of Flexner returning in the fall and launching a vengeful campaign against his enemies.[58]

For some time, Earle had expected Maass to initiate a consultation. Still waiting on July 21, Earle could stand it no longer. He compiled a damning collection of allegations against Flexner, and sent them to Leidesdorf and Maass. The main thrust of the complaint involved grants that Earle was seeking from the Rockefeller Foundation and Carnegie Corporation. The requested funds were to support three visiting scholars for a seminar on American Foreign Relations. Earle claimed that Flexner had spitefully sabotaged the Rockefeller application. "Discretion requires that I write nothing of the details, although I could discuss them verbally with you if you wish."[59]

Earle had overcome his earlier preoccupation with propriety. The attacks, in his cover letter and statement, were libelous. Not only did he warn that trustee inaction could lead to Einstein's resignation, but Earle implied that anti-Semitism was behind Flexner's treatment of one of the proposed visitors. In touting the merits of his three scholars, Earle compared them to two other candidates who were allocated *Institute* funds. Both of these men were described as Flexner family friends who had been opposed by the faculty.

One day after writing to Maass, Earle received information from the Rockefeller Foundation that led him to partially recant the charges against Flexner. Even so, the impact of Earle's memo may have lain more in its tone than substance. Any reader had to be struck by the bitter feelings it exposed of some faculty toward Flexner. Maass immediately requested Earle's permission to show the statement to Hardin.[60]

Hardin had recently emerged as a mediator. In many respects he was better positioned, than Aydelotte, to reach Flexner. The relationship between Flexner and the younger Aydelotte was that of mentor to protegé. In contrast, Flexner viewed the 79-year-old Hardin as a senior advisor and as Bamberger's emissary. Moreover, Hardin held the unique status of being a trustee of both the Institute and Princeton University. Flexner was disposed to listen to Hardin and pontificate to Aydelotte.

Hardin took his turn at persuading Flexner. The conversation could have begun on their common ground that Maass was a jerk and the founders were saints. Beyond this, Hardin was diplomatic but firm. He told Flexner that the time for retirement was at hand. Hardin warned that undue delay could result in unpleasantness and be hurtful to Bamberger. It is unclear whether Hardin went so far as to say that the founders favored retirement.[61]

When Hardin left Flexner, it was not immediately clear whether the conversation had produced its intended effect. Hardin then turned to heading off Leidesdorf's and Maass' proposal for an emergency Board meeting. To Hardin, a diplomatic solution still seemed within reach, but he knew time was running out. It was late July. From Earle's letters, it was apparent that the controversy could not be contained much longer. Too many people knew about the problems, and ugly gossip would soon spread. Hardin wrote to Flexner seeking to learn his plans. Leidesdorf and Maass agreed to wait for the answer.[62]

It would take several days for Flexner to respond. He had decided to discuss his future with Bamberger and Fuld, who were vacationing at Lake Placid. The only portrayal of these consultations is embodied in Flexner's July 29 reply to Hardin.

I discussed with both Mr. Bamberger and Mrs. Fuld, separately and together, the question of my retirement and the manner in which it should
be managed . . . . While they regret that the passage of time makes a step
of this kind *necessary and inevitable* [emphasis added], they wish me to do
anything which I myself prefer and in the way I prefer.[63]

Flexner went on to quote Bamberger insisting, "You must retire in a dignified
and seemly way . . . so that no one can possibly say that you have been
pushed out."

Despite his penchant for filtering out the negative, Flexner was admitting that the founders expected him to retire. The message had finally got-
ten through, at least to some degree. Further information about Flexner's
state of mind was provided in a letter from Mrs. Flexner to Mrs. Aydelotte.
Mrs. Flexner described her husband, after returning from Lake Placid, as
"unhappy" and "the prey of different ideas."[64] Apparently, Flexner's powers
of denial were not entirely defeated. This may explain, partly, his failure to
appreciate the urgency of action. No timetable was included in the letter to
Hardin. Flexner merely indicated that his retirement would remain confidential until the trustees were informed at the October meeting.[65]

To Flexner's opponents, the letter appeared to be another stalling tactic.
Leidesdorf, Maass, and Aydelotte resumed their call for an emergency meeting
of the Board. Even Hardin was somewhat frustrated. His reply to Flexner
went straight to the point, noting that Flexner's letter was "rather cryptic as
to the real nature of your plans, and I cannot but feel that it would have been
better to have anticipated possible unfriendly action by pursuing the course I
outlined in our last interview."[66]

By then it was early August. The factions were positioning themselves
for the endgame. While Hardin pointed to his recent progress and pleaded
for more time, Leidesdorf and Maass prepared to travel to Lake Placid, where
they would make a direct appeal to Bamberger. The faculty were also active.
In their correspondence, Earle and Veblen had each been asking the other
to bring Einstein into the campaign. Veblen finally took on the task. He
prepared a chronology of the summer developments and sent it to Einstein at
Long Island. Veblen asked Einstein to communicate his views to Leidesdorf
and Maass.[67]

Einstein was accustomed to receiving appeals for support. Veblen's request arrived just after Einstein was lobbied by physicist Leo Szilard on another matter. Recent scientific developments had opened the possibility of applying nuclear physics to the design of powerful bombs. Szilard was alarmed

at the prospect of Hitler gaining access to this technology. Szilard urged Einstein to exert his influence by alerting certain international leaders to the danger. The outcome was an historic letter to President Roosevelt that was signed by Einstein.[68] The draft was provided by Szilard and dated August 2.*

The Flexner coup was another cause that Einstein endorsed. For some time he had stood among the dissident faculty. Einstein was willing to join in persuading Leidesdorf and Maass, but it was natural for him to ask that Veblen compose the words. When Veblen received the request, it had already been mooted by fast-breaking developments.[69]

The meeting with the founders had taken its toll on Flexner. For the past decade he had operated with a self-assuredness anchored by two core beliefs: that he was an outstanding director who served with Bamberger's mandate. To learn otherwise was a devastating blow. In letters to his daughter Jean, Flexner remained upbeat. He spoke of *his decision* to retire and of the welcome freedom it would provide. Meanwhile, he was sinking into a deep depression.[70]

In the midst of this disturbed state, Flexner received Hardin's letter containing the comment about cryptic plans. Over the next few days Flexner reacted erratically. A frequent impulse was to resign immediately.[71] In the end Anne prevailed on her husband to place himself in Hardin's hands. On August 7 Flexner wrote Hardin asking him to "give me your judgment as to the date on which I should retire."[72]

Hardin's reply arrived on August 12.[73] Later that day Flexner wrote to the trustees. With the first sentence he finally followed Hardin's advice: "I take this means to inform you that I shall, at the meeting of the Board to be held October 9, ask to be relieved of my duties as Director of the Institute, effective as of that date."[74]

---

*See [Clark, *Einstein*] for more details and the letter.

Robert Oppenheimer and John von Neumann in front of the IAS computer, ca. 1952. (Alan Richards photographer. Courtesy of the Archives of the Institute for Advanced Study. Printed by permission of Bernice Sheasley.)

CHAPTER TEN

FAST FORWARD

When Bamberger, Fuld, and Hardin joined Maass and Leidesdorf in seeking a change in director, Flexner's days became numbered. He could not win. His struggle continued until the August 12, 1939, letter, from which there was no turning back. Flexner's retirement was on the record. The toll on the 72-year-old director was considerable. In the night after mailing his notification to the trustees, Flexner was awakened by severe chest pains. Mrs. Flexner was terrified that her husband was having a heart attack. Medical help was summoned. Without any electro-cardiogram equipment in the vicinity, the doctor prescribed extended bed rest. No further heart episodes occurred, but Flexner was exhausted and miserable. He obsessed over an enemy list headed by Maass and Veblen.[1]

Flexner set himself to managing the transfer of power. He was determined to install Aydelotte as his successor, denying, at all costs, the position to Veblen. Having severed his connections to various key trustees, Flexner was limited in his capacity for orchestrating this last act. With Aydelotte then in Mexico, Flexner prevailed upon Walter Stewart to assist him. Stewart reached the Canadian camp on August 22, where he listened to Flexner vent and plot. A few days later Aydelotte returned to Swarthmore, only to find himself inundated with intrigue from all sides.[2]

In principle, only the trustees were informed of Flexner's decision to retire. A crucial question was whether the Board would take up the issue of a new director prior to its regularly scheduled meeting in October. The availability of the Board Chairman, Alanson Houghton, was in doubt due to a recent strenuous trip and his own failing health. Flexner feared the scenario of a special meeting conducted in the absence of Houghton and other trustees who might be unable to attend on short notice. The Vice Chair Maass would preside, placing him in a position to steer support to Veblen. For their part,

Maass and Leidesdorf were worried that any extraordinary session might compromise the confidentiality of Flexner's retirement, giving him an excuse to renege. Earle, who was informed by both Maass and Veblen, first thought that Flexner's resignation letter was another stalling ploy. In any event Earle decided that it was important for the faculty to assert its role in the selection of a new director. Earle lobbied Maass, Veblen, and Einstein without result.[3]

Aydelotte was privy to all of these views, as well as to the knowledge that he was Bamberger's choice to succeed Flexner. That Aydelotte was the central communication node attested to his unique standing in the shattered Institute family. Nobody was more attuned to the feuds or better positioned to bring about some healing. Aydelotte was an obvious choice to become the new director. After nearly two decades at Swarthmore, the 58-year-old college president viewed the opportunity as an ideal final stage of a career devoted to scholarship. Aydelotte was excited at the prospect of implementing his own ideas on subjects, such as Latin-American studies, for Institute development.[4]

The one call for quick trustee action came from Aydelotte. He knew that the Institute would need him to replace Flexner immediately. In order that the search for a Swarthmore successor begin as soon as possible, Aydelotte was anxious to present his resignation to the Swarthmore Board at its mid-September gathering. It was not to be. Flexner was too paranoid to permit an early Institute meeting. The fears were unfounded. Whether or not Veblen had designs on the position, Bamberger was set on Aydelotte. The outcome of any trustee deliberation was determined, regardless of when it was held and who attended.[5]

Early in October Bamberger provided Houghton with the script for the Institute Board meeting. As a concession to Aydelotte, Houghton went to Princeton early and consulted the faculty for their views on director candidates. Most of the trustees were present for the meeting on Monday, October 9. When the time came, Houghton appointed Bamberger's preselected special committee to nominate a new director. The committee consisted of Hardin, Leidesdorf, Bamberger, and Houghton. The Board recessed while the committee deliberated and then returned with their recommendation of Aydelotte.[6]

It was at this point that a brief departure from the script occurred. Percy Straus attempted to initiate a discussion on whether someone younger than Aydelotte should be chosen for the position. Taken aback, Houghton abruptly cut off Straus and moved the meeting back on track. Aydelotte was elected as director.[7]

A few days later Straus resigned from the Board, ending a notable tenure. Straus was from the prominent family that controlled Macy's department store. He was regarded by Houghton and Flexner as a typical ornamental trustee. Yet Straus had taken his role seriously, thinking about issues and periodically expressing concerns to Flexner and his colleagues. On a Board that rarely pondered matters of substance, it was Straus who warned about the budget crunch that loomed with the start-up of the third school. In bringing up Aydelotte's age, Straus anticipated, by many years, the next crisis that would face the trustees.[8]

Flexner left Princeton in the middle of November, heading south for rest and treatment. In his parting message to Aydelotte, Flexner candidly revealed his long held insecurity:

> you had one great advantage over me—you are in your own right a scholar and can be one of the humanistic group. I, alas, have never been a scholar, for two years at the Johns Hopkins between 1884 and 1886 do not produce scholarship, though they do produce and did produce a reverence for it which I am now leaving in safe keeping with you.[9]

As it would turn out, Aydelotte's real advantage was that he understood what was important to scholars, and in which administrative matters they wanted to participate. The overcompensating Flexner never did. As soon as Flexner left, Aydelotte called together the faculty, explaining that there would be two or three meetings a year. To improve communication and increase transparency, Aydelotte established a standing advisory committee of three professors, one from each school. These good faith measures boosted the morale of a faculty that had felt marginalized under Flexner.[10]

It was fortunate that Aydelotte was familiar with the Institute's personnel and its operation. For, until a successor was in place at Swarthmore, he served as chief executive of both institutions, spending Thursday, Friday, and Saturday in Princeton. Among his first moves was to draft a resolution of tribute to Flexner and see that the outgoing director received a dignified severance. An initial look at the budget revealed the gap that was still unfilled by the Rockefeller Foundation. If Aydelotte had examined the numbers more closely, he might not have been so generous as to approve a $4,000 raise in his predecessor's pension to $12,000.[11]

Aydelotte had been a trustee since the Institute's founding. He thought that he had conscientiously fulfilled his role, attending meetings, serving on committees, and reading reports. Yet it came as quite a shock when Aydelotte did a close reading of the budget. There was no provision for the operation of Fuld Hall nor for stocking its library with books. Together with Flexner's pension, the salaries of the new economists, and an already projected deficit, the Institute was seriously in the red. Aydelotte's plans for new programs in Latin-American studies and Chinese studies would have to be deferred until fiscal order could be restored.[12]

Discussing the problems with Bamberger yielded limited results. Not only did Bamberger feel that his previous contributions provided a more than adequate endowment for the Institute, but he was considering other causes for his remaining fortune. Still, Bamberger was sympathetic to the predicament of his new director, agreeing to contribute $25,000 for each of the next four years toward the purchase of library books.[13]

In the spring of 1940, the Rockefeller Foundation approved a three-year matching grant to the Institute for the program in economics. Both parties were to contribute $35,000 per year. The unusual manner in which Aydelotte negotiated the arrangement revealed the extent to which Bamberger had distanced himself from providing further support. It was only after the grant was formally awarded that Aydelotte asked Bamberger to donate the Institute share. Bamberger supplied the matching funds, and other agencies awarded smaller targeted grants. The books appeared to be balanced, but Aydelotte recognized that the picture was illusory. The Institute was relying on too much soft money to finance long-term commitments.[14]

Over his first year as director, Aydelotte made substantial progress in cleaning up the messes left behind by Flexner. Discontent among the faculty abated and temporary control was gained over the budget. During this period Aydelotte took his regular salary from Swarthmore and received no compensation from the Institute. The next year a new president took over at Swarthmore, and Aydelotte went on the Institute payroll.

The budget problems were terribly disconcerting to Aydelotte. He had come to the Institute with the expectation of building on the work of Flexner. The most appealing aspect of being director was the prospect of leading new directions for development. The difficulty was that all of the $320,000 endowment income and $115,000 in grants were required to meet ongoing expenses. Even to achieve this balance, the outlay for members had been cut to an unacceptably low level. Moreover, the most predictable upcoming budget forces were likely to be exerted in the wrong direction. Certain faculty would merit salary raises and the grant income would sunset. Unlike a university president, Aydelotte lacked the option of seeking an increase in tuition. The

Institute's continued viability depended on Bamberger contributing millions more to the endowment, including, at the very least, capitalization of the $115,000 shortfall. Otherwise, Aydelotte projected that he would be forced to preside over cutbacks rather than enhancements.[15]

Aydelotte made these points in a forthright November 1940 memorandum to Bamberger and Fuld. The founders were unresponsive. However, about this time Aydelotte and Bamberger began to meet regularly for lunch. Each week Aydelotte made his case and Bamberger listened patiently. The pitch for money was no less subtle that it had been from Flexner. The biggest difference in the style of the new director was his genuine humility. Bamberger must have been relieved to contemplate his Institute without the interminable name dropping and self-serving comparisons to Johns Hopkins and the Rockefeller Institute.[16]

The luncheons continued over two years with Aydelotte continually hoping for some sign from the inscrutable Bamberger. During this period the grants neared their end and the faculty grayed. Einstein, Veblen, Lowe, and Herzfeld were approaching the retirement age of 65. The annuities afforded by their pension fund accumulations were each under $3,000. The payouts would be nowhere near the $8,000 guaranteed to Veblen and Einstein nor even be at subsistence level for Lowe and Herzfeld. Aydelotte felt that the Institute had a moral obligation to supplement the pensions of Lowe and Herzfeld. The prospect put additional stress on a budget for which the long-term projections were already frightening.[17]

Late in 1942, Aydelotte decided that the time had come to ask Bamberger about his intentions. To prepare the ground, he drafted a memorandum that began with the assertion that Flexner's vision had been validated. Aydelotte went on to claim that the Institute was a unique and vital resource in American education and scholarship, but to remain on the cutting edge required the cultivation of new subjects. The current $8,000,000 endowment was inadequate to support existing programs.

> We have made a good beginning, but it is only a beginning. If eventually we have a larger endowment we can realize the possibilities of this unique educational enterprise—if not, we shall be compelled to curtail our activities and to disappoint the expectations of scholars all over.[18]

When Bamberger requested elaboration on possible new directions, Aydelotte followed up with another memorandum proposing the establishment of Schools of Oriental and Latin-American Studies along with a temporary program in English literature.[19]

Unbeknownst to Aydelotte, on February 20, 1943, Bamberger filed what would become his last will and testament. The Institute was the residual legatee. Perhaps Bamberger wanted to preserve the flexibility to make further amendments because he did not immediately inform his anxious director. Bamberger was then 87 and living amidst the uncertainty of World War II. As the year went on Aydelotte did notice a softening of the founder's attitude in discussing the budget problems. Reassurance replaced guardedness. At some point Bamberger became more explicit about his commitment, but he never permitted Aydelotte to move forward on a new school.[20]

The Board did take up the pension problems. These complex issues were addressed throughout 1943. Flexner's administrative shortsightedness was illustrated by the circumstances of the professors slated to retire over the next two years. No budgeting provision had been made to finance the $8,000 guarantees to Einstein and Veblen. The retirement stipends to Herzfeld and Lowe were inadequate. During the summer a plan was adopted to subsidize the stipends to Lowe and Herzfeld while jointly increasing contributions for other professors so as to assure that the pensions of all faculty, excepting Goldman, reached the $4,000 ballpark. The added expense for the Institute left no doubt that Aydelotte's ambitions for development were subordinate to his interest in the welfare of the faculty.[21]

When Lowe and Herzfeld learned of the arrangement, they protested. Both men were engaged in projects that were years from completion. Their understanding from Flexner was that they were to be exempted from the normal retirement age. Then Einstein and Veblen expressed a desire to remain on regular salary. Aydelotte faced another problem. Waivers of the mandatory retirement age had serious implications for the budget. Aydelotte, who had yet to hire a single new professor, had already given up hope of replacing retiring faculty. However, he was counting on the reduced financial obligation to retirees in making up for the soft money lost from expiring grants.[22]

The ultimate resolution was to allow exceptions to the retirement age for Einstein and Veblen, owing to "their distinguished service to the Institute."[23] Lowe and Herzfeld were forced to retire on schedule. In addition to augmenting the humanistic studies professors' annuities to $4,000, special funds were established to support Lowe and Herzfeld in their continuing research and publication. In response to a request from Aydelotte, Bamberger agreed to contribute $14,000 each year to make up for the difference between the pensions and salaries of Einstein and Veblen. Thus it would be represented that there was no adverse impact on the budget from the special treatment accorded to Einstein and Veblen. Even so, Aydelotte was still looking at contingency plans to reduce salaries a couple of years down the road.[24]

A sad milestone in Institute history was reached on March 11, 1944, when Louis Bamberger passed away. He was survived by his sister Carrie Fuld for just four months. Their estates would double the Institute's $8,000,000 endowment. A majority of the original trustees were gone, leaving Aydelotte, Flexner, Maass, Leidesdorf, Hardin, Weed, and Edgar Bamberger from the charter group. When Houghton died in 1941 the formal leadership roles were divided between Hardin as chairman and Maass as president, but Louis Bamberger continued to make the important decisions. Although Hardin presided over trustee meetings, he acted as Bamberger's representative, rarely moving on his own initiative. Bamberger's death left a large power vacuum on the Board. Maass, who tended to push his own opinions, wasted no time in moving to consolidate his authority.

The April Board meeting took place between the deaths of Bamberger and Fuld. Maass arrived with a formal motion for the establishment of a committee "to make a survey of what has been done by the Institute in the past, what is being done at present, and of plans for future development."[25] The suggestion was sensible and seemingly innocuous. With Bamberger's death, the turnover on the Board, and the infusion of new funds, it was an appropriate time to pause and take stock. Naturally, Maass became the chairman of this Committee on Institute Policy that also included Leidesdorf and three newer trustees. It was notable that Aydelotte was excluded from even ex-officio status.

Maass then revealed his true agenda. He contacted Aydelotte to inquire whether the director intended to retire when he reached 65 the following year. Mandatory retirement of the director was not covered by Institute by-laws. The information was relevant to the report of the Committee on Institute Policy, but not seemingly an issue on which the Committee was charged to make a judgment. The question had been considered earlier by Aydelotte who had concluded, on principle, that it was appropriate for the director to be bound by the same retirement regulations that he enforced on the faculty. Aydelotte had expressed the opinion to Bamberger, who disagreed, specifically requesting Aydelotte to stay on beyond the age of 65. Now Bamberger was out of the picture and the influential president of the trustees appeared, by the very fact that he raised the issue, to favor retirement. Aydelotte gave considerable thought to his response.[26]

Maass' later actions would confirm that he was using the Committee on Institute Policy as an instrument to force Aydelotte out of office. It is a mys-

From left to right: James Alexander, Albert Einstein (holding chalk), Frank Aydelotte, Oswald Veblen, Marston Moore, in the mid 1940s. (Courtesy of the Archives of the Institute for Advanced Study.)

tery as to what prompted Maass to take this drastic step. Aydelotte had performed his duties admirably, especially considering the hand he was dealt. No record exists of his having any conflict or dispute with Maass. There was one trustee actively campaigning against Aydelotte, but his views were unlikely to be persuasive with Maass. Flexner had turned against Aydelotte when the decision was made to permit Einstein and Veblen to continue on full salary.[27]

In October 1944 Aydelotte informed the Committee of his intention to retire at the age of 65. However, two subsequent developments, involving Flexner and the faculty, soon caused Aydelotte to reconsider. Flexner injected himself into a bitter feud between humanistic studies professor Erwin Panofsky and member Charles de Tolnay. Flexner had a high opinion of de Tolnay and his work on Michelangelo. Although Flexner was uninformed as to the particulars of the controversy, he took the side of de Tolnay, threatening Panofsky with dismissal from the faculty. It was an ugly episode that led Aydelotte to chastise Flexner for improper interference. At Flexner's insistence the case went to the trustees.[28]

With Aydelotte feeling that his administration was under attack, the second development quickly brought out his backers. When the faculty learned of the retirement plans, they petitioned the trustees. The resolution, supported by all of the professors except Stewart and Warren, urged that the current director remain on beyond the age of 65. Aydelotte was gratified. He felt a stronger attachment to the faculty than to the Board. The arguments for retirement had become clouded. Now Aydelotte was inclined to stay on, out of loyalty, rather than depart out of principle.[29]

The faculty involvement angered Maass. He cared little about the views of the professors, but appreciated the advantages of a smooth abdication. In January 1945 the Committee on Institute Policy reported a compromise recommendation. The proposal was for Aydelotte to retire at 67 with a pension sweetened to the $12,000 level of Flexner. The Committee was hoping that Aydelotte would accept the plan, eliminating the possibility of another power struggle. The question would be called at the March Board meeting.[30]

Aydelotte was torn over how to respond. In the end he abstained, absenting himself from the Board meeting and leaving his fate in the hands of the trustees. Aydelotte's nondeclaration set up a contentious March showdown between Maass and Veblen. The former allies, in overthrowing Flexner, engaged in a heated argument on whether to force Aydelotte into retirement. Veblen's case on behalf of Aydelotte was supported by the founders' nephews Edgar Bamberger and Michael Schaap (whom his uncle brought onto the Board in 1941). Both nephews attempted to represent the wishes of their late relatives. It was not enough. Maass was joined by Leidesdorf, Hardin, and four newer trustees. Flexner did not attend.[31]

The outcome set up a redistribution of power. Aydelotte was a diminished two-year lame duck who was more or less reporting to Maass. The next order of business was to hire new faculty. Neither Maass nor Aydelotte was satisfied with Flexner's recruiting practices. Maass objected to serving as a rubber stamp while Aydelotte felt that faculty participation should be an essential ingredient. They worked out the general parameters of a new procedure. Nominations were to originate within schools and be subject to approval by the entire faculty. The director would then submit names to a new standing committee of the trustees, the Committee on Appointments, which would review the suggestions prior to consideration by the Board. In effect the director would act as a first filter and the Committee on Appointments the second. In explaining the procedure to the faculty Aydelotte promised that, in making his recommendations to the Board, prior approval by the faculty was a necessary, but not sufficient, criterion. Maass was appointed to chair the Committee on Appointments, with Aydelotte and Weed selected as members.[32]

The process moved quickly. For years Aydelotte and the faculty had discussed hypothetical appointments. With the funds finally available, everyone was ready with candidates. The School of Economics and Politics put forward the name of Jacob Viner. Viner was the economist who had been originally recommended to Flexner to lead the School. That, over fourteen years, Viner had maintained his high esteem reflected on both the economist and the director who had adjudged him to be insufficiently "polished."[33]

The School of Mathematics nominated two brilliant scientists, physicist Wolfgang Pauli and mathematician Carl Ludwig Siegel. Both had been in residence since 1940. They were among the many European refugees whom Veblen and the Institute assisted in relocating during the War period. The choice of Humanistic Studies was William Albright, a Johns Hopkins archeologist whose expertise on the Near East would fill the void left by the retiring Herzfeld.

Individual cases were presented for faculty votes at a meeting in May 1945. Each name was introduced by a designated school representative who highlighted the candidate's accomplishments and quoted endorsements from distinguished outside scholars. Under an experiment intended to promote understanding between faculty and trustees, Maass and Weed attended the session. As the nominations proceeded, Maass assumed the role of devil's advocate, challenging every appointment. The chasm between Maass and the faculty gradually became apparent. The faculty were basing their considerations on scholarly substance while Maass expected some business-type rationale to justify investment of Institute resources in one school over another. He did not appreciate the concept of basic research. The discussion took on an adversarial tone, reaching its low point when Maass belligerently served notice of his intention to assume control over the hiring process.[34]

Despite the disastrous interaction of Maass with the faculty, the trustees authorized Aydelotte to move forward on Viner, Pauli, and Siegel. Several complications then arose. The offer to Viner was withheld when it was learned that he was under recruitment by Princeton University. The School of Economics and Politics then received approval for the Harvard maritime historian Samuel Eliot Morison, who subsequently declined. Pauli had left the ETH in Zurich with the understanding that his professorship was to be held open until it was safe for him to go back after the war. Under these circumstances Pauli decided it was improper to consider a permanent Institute appointment. He accepted a visiting professorship. A few months later Pauli was awarded the Nobel Prize in physics. The following year he returned to Zurich. Siegel did accept an Institute professorship, but constantly felt out of place in America. Although the mathematicians strived to accommodate him, Siegel eventually resigned and completed his career at Göttingen.[35]

In addition to Pauli and Siegel, the School of Mathematics moved to regularize another personnel arrangement. Kurt Gödel had been a member intermittently since 1933 and continuously from 1940. Gödel unquestionably possessed the theorems for an Institute professorship. His foundational work had profoundly influenced the perspective of mathematicians. When the financial situation improved in 1945, von Neumann and Veblen decided it was time to put the relationship on a permanent basis. Yet Gödel's mental instability and antisocial behavior raised questions about his suitability for a faculty position.[36]

After securing Aydelotte's approval, Veblen went before the faculty to propose Gödel for appointment as a "permanent member." The reaction to Gödel's nomination for this new intermediate standing was a bit surprising. Unlike frequent turf battles where candidates are impeached by professors from other schools, Stewart and Lowe suggested upgrading the consideration to a professorship. The minutes of the meeting state, without specifying names: "This was rejected by the mathematicians." [37] In 1946 Gödel became a permanent member at a salary of $6,000. Seven years later he would be made a professor.

The entire first generation of Institute mathematics faculty remained in Fuld Hall, but the years had lessened their vitality. Weyl had passed 60. Alexander's last important work, on cohomology, was a decade old. He started to withdraw from Institute activity, requesting leave and then a reduction to permanent member status.[38] It was in Morse and von Neumann that the energy of the mathematics faculty was concentrated. Fortunately, both professors were more than amply endowed with this resource. However, in the summer of 1945, von Neumann began exploring other situations which might be better suited to pursuing his passion for computers.*

During the war von Neumann became increasingly involved in weapons research. Some indication of how this affected his mathematical predilections

---

*For a more detailed history of these developments see *John von Neumann and the Origins of Modern Computing* by William Aspray.

was given in a 1943 letter where von Neumann commented that he had "developed an obscene interest in computational techniques."[39] Next came his participation in the Manhattan Project. At Los Alamos von Neumann recognized the need for more powerful computational devices. His consultations with scientists and engineers at the University of Pennsylvania stimulated von Neumann's thinking on the design of electronic computers. In 1945 von Neumann decided to launch a major project to construct a stored program computer.

With the pervasive role of computers in society today, it is difficult, now, to appreciate how the proposal appeared in 1945. Certainly the engineering-laboratory nature of the work was unprecedented for the Institute, as were the costs. The initial budget estimate totaled $300,000 over several years. Implementation required staff and a new building. While von Neumann hoped to pursue the work at the Institute, he realized that it might be necessary to go elsewhere. Thus von Neumann opened discussions with MIT and the University of Chicago as he followed channels at the Institute. Both MIT and Chicago responded with offers.[40]

The reaction of the Institute mathematics faculty to the project was, at best, mixed. From the mathematical perspective, there was no enthusiasm for the problems. Yet it was clear that von Neumann and the computer were bundled. Since von Neumann was widely viewed as such a vital member of the faculty, the mathematicians would not stand in his way. The biggest obstruction remained funding. Prospects did exist for obtaining outside support. To maximize these opportunities it would be desirable for the Institute to front the project with $100,000.[41]

Aydelotte was faced with whether to recommend a major allocation to the trustees. Two years earlier it would have been impossible. Now it could be obtained out of the income added from the Bamberger and Fuld bequests. Programming such a large sum for the electronic computer project was likely to jeopardize Aydelotte's chances of moving on Chinese studies and other schools. Aydelotte wisely decided to back von Neumann and the computer. The Board voted to appropriate $100,000 from its surplus income and von Neumann remained at the Institute.[42]

It would be the one major initiative of Aydelotte's regime. Shortly after obtaining approval, he answered a call from President Truman. During the first several months of 1946 Aydelotte served full-time on a joint Anglo-American commission to make recommendations on the future of Palestine. Over that interval the Institute was administered by a faculty committee headed by Marston Morse. When Aydelotte returned for his final year as Institute director, he attempted to start a program in Chinese studies. The Board decided to defer a decision until the arrival of Aydelotte's successor. Two significant personnel moves were permitted. Classical archeologist Homer Thompson accepted a professorship in the School of Humanistic Studies. The 28-year-old physicist Richard Feynman declined a joint offer to serve on the faculty of the university and as a permanent member of the Institute.[43]

In looking back on Frank Aydelotte's eight-year tenure as director, there are different interpretations of his legacy. Aydelotte greenlighted the computer project that arguably led to one of the greatest scientific developments of the century. The offers to Pauli, Siegel, Morison, Gödel, Feynman, and Thompson set an impressively high standard of scholarship. However, of these only Gödel and Thompson remained at the Institute, and no new schools were created. Any such statistical or anecdotal analysis misses Aydelotte's true contribution. He stabilized an institution that was in serious jeopardy. It is impossible to determine how Bamberger and Fuld would have disposed of their estates under a different director. What is certain is that Aydelotte handed over a directorship with better prospects than the administrative post he himself had assumed.

There was plenty of time to select Aydelotte's successor. At the April 1945 trustees meeting, Hardin appointed a search committee chaired by one of the newer trustees, Henry Moe of the Guggenheim Foundation. Another personnel move, taken at the same meeting, would have a significant impact on the choice of a new director. Lewis Strauss was elected to join the Institute Board. An opening had developed when Flexner agreed to step aside and make room for a younger person.[44]

The fifty-year-old Strauss* was ambitious, wealthy and well connected. With only a high school education he had become a special assistant to the Secretary of the Navy and was a partner in a major Wall Street investment firm. It was expected that Strauss would be helpful to Leidesdorf in managing the Institute's portfolio.[45]

As was characteristic of Strauss' career, he quickly became an influential member of the Institute Board. When Hardin died at the end of 1945, Maass added the title of chairman to the presidency he already held. The rookie Strauss became vice chairman of the Board. Meanwhile, little was happening

---

*Note the extra s from the spelling of Percy Straus' last name.

with the search for a new director. In October 1946 Moe resigned from participation in Institute activities. Maass appointed Strauss to take over as chair of the search committee. Later that month President Truman named Strauss to serve on the newly established Atomic Energy Commission (AEC). By law the AEC was to have control over the United States' atomic weapons, resources, and policy. Strauss was one of just five commissioners.[46]

Despite the substantial demands of the AEC post, Strauss moved to identify a new director for the Institute. Rather than hold meetings, Strauss consulted individually with faculty and search committee members. By this approach Strauss maintained control over who received serious consideration. The faculty had previously produced a list of nominations that included Detlev Bronk, Robert Oppenheimer, and Linus Pauling. The name of Oppenheimer had a particular resonance with Strauss. It was not just because Oppenheimer was the famous scientist-administrator who had so successfully directed the Manhattan Project. At the same time Strauss was selected for the AEC, Oppenheimer was one of the scientists appointed by President Truman to the General Advisory Committee, set up to provide technical advice and expertise to the AEC.[47]

Strauss and Oppenheimer were not well acquainted. They had met only once, at a conference a year earlier in the office of the Secretary of War. Observing Oppenheimer in such a setting was enough to impress most people. His erudition, articulate command of nuclear issues, and piercing blue eyes always dominated these discussions. On paper, Oppenheimer had impressive credentials to direct the Institute. In addition to his leadership of the Manhattan Project, the 42-year-old professor had lifted the Berkeley theoretical physics program to the highest standing in America. Strauss and the search committee decided that Oppenheimer was their first choice to replace Aydelotte.[48]

Strauss offered the position to Oppenheimer when AEC matters brought the two together in San Francisco. The Institute proposition intrigued Oppenheimer. After the Manhattan Project, he had had difficulty returning to his life as a physics professor at Caltech and Berkeley. The old routine was incompatible with his new role as the world's foremost authority on atomic energy. Oppenheimer was constantly summoned to Washington for consultations. On the logistical level alone, Princeton's location on the east coast train route offered an appealing respite from the exhausting transcontinental air travel.

Other factors made the Institute a good fit for Oppenheimer. His intellectual interests went far beyond physics. He read Greek, Latin, and Sanskrit.

Oppenheimer's knowledge spanned literature, history, and a vast range of subjects. The Institute directorship offered him the opportunity to do physics, set an agenda for high level research, and continue his involvement in government policy. After delaying for several months, Oppenheimer accepted the job.

Oppenheimer arrived in Princeton during the summer of 1947. He quickly sized up the state of the three schools and their faculties. Mathematics was "healthy and flourishing" and the new director was confident that he would build strength in physics. As for the other two schools, Oppenheimer was negative, pointing in particular to "a certain insularity of their efforts."[49]

The School of Economics and Politics was in serious trouble. There were internal problems and questions about the suitability of its personnel. The political scientist Mitrany had been in England on leave since the start of the War in 1939. Through their correspondence Mitrany and Earle had come to despise each other. On one judgment they were in agreement. Both had little regard for the scholarship of the Institute economists, a view shared by Oppenheimer. Of the school's five professors, only Earle was meeting expectations.[50]

In the School of Humanistic Studies Goldman had just retired. The remaining faculty of Meritt, Panofsky, and Thompson was outstanding. Yet Oppenheimer pointed out that in the domain of Humanistic Studies "there are obviously areas of great fruitfulness beyond the Hellenistic studies to which the Institute is already committed."[51] He immediately moved to broaden the program by bringing in the classical philosophy scholar Harold Cherniss.

With his enormous self-confidence, Oppenheimer was accustomed to acting decisively. He quickly put an end to consideration of Chinese Studies, explaining "there is no one outstanding man available for such an undertaking." This did not mean that other schools were not feasible. To explore these possibilities Oppenheimer asked the trustees for a $120,000 special fund over a five-year period. The Board approved what became known as the Director's Fund with which Oppenheimer had discretion to select scholars in areas outside those of the current faculty.[52]

A stunning transformation immediately took place in theoretical physics. Practically overnight the Institute became a world center for the ventilation of ground breaking developments. Among the luminaries in residence for at least a term of 1948 were Niels Bohr, Hideki Yukawa, George Uhlenbeck, and Paul Dirac. The most promising young physicists were invited as members.

Meanwhile the School of Mathematics was fortified with permanent members on track to become faculty. This second generation included physicist Abraham Pais and mathematicians Deane Montgomery and Atle Selberg.

As mathematics and physics flourished, Oppenheimer set about reconfiguring the other two schools. An opportunity to phase out economics arose when Winfield Riefler accepted a position in Washington at the Federal Reserve Board. With Walter Stewart approaching retirement, the only remaining economics professor would be Robert Warren. The downside of these changes was the imbalance it would create among the Institute's schools. Oppenheimer persuaded the School of Economics and Politics to merge with Humanistic Studies, pointing to their common ground. In 1949 they formed the School of Historical Studies. One year later Stewart retired and Warren died, closing out the Flexner foray into economics.[53]

Oppenheimer's presence brought an incongruous sight to the Institute's bucolic setting. Situated outside the director's office were a large safe and an armed guard. These security arrangements protected the secret documents that were at Oppenheimer's disposal as the chair of the General Advisory Committee. As he was putting his imprint on the Institute, the father of the atomic bomb attempted to shape the United States' policy on atomic energy. These latter efforts were the source of considerable frustration.

Oppenheimer was a complex person for whom leadership of the Manhattan Project had held out several appealing prospects. He could do his part to stop Hitler while gaining notoriety and leading some intriguing science. The three years of intense work had culminated in a spectacular achievement of the scientific goals. Yet the bomb decimated Japanese civilians rather than Hitler. As time went on Oppenheimer became increasingly conflicted between the pride in his accomplishment and guilt over the outcome. He began to question whether the use of the bomb, in which he had acquiesced, was really justifiable.[54]

What was clear to Oppenheimer was that the atomic bomb was a devastating weapon for which there was no defense. Proliferation and use were inevitable. Seeing no other remedy, Oppenheimer concluded that the welfare of the world hinged on the establishment of international control over atomic power. His proposal called upon the United States to share knowledge, relinquishing its strategic advantage over the Soviets. It was a tough sell to make to American politicians at the outset of the Cold War. Oppenheimer lost the battle, but continued to work within the government for a more open policy.

Among the opponents of international cooperation was AEC and Institute Board member Lewis Strauss. To Strauss the world was in an ongoing struggle between the good and evil forces of the United States and Soviet Communism. Survival depended on maintaining military superiority. Suddenly, in 1947, Strauss and Oppenheimer were bound together to set policy on atomic energy and for the Institute. With their arrogant personalities and different world views, it was a collaboration destined for trouble. Matters intensified over the summer of 1949. In May Strauss became president of the Institute Board, effectively second in command behind the chairman Maass. One month later Oppenheimer and Strauss testified before a Congressional Committee that was investigating the operation of the AEC.[55]

Strauss went first. He explained his dissent from the AEC's approval for the export of radioisotopes to Norway. The security concerns expressed by Strauss were then rebutted by Oppenheimer in a manner that elicited laughter from the audience. Strauss sat humiliated.[56] Over the next few years Oppenheimer and Strauss disagreed over fundamental issues such as the development of the hydrogen bomb. There were other disputes on Institute matters.[57]

When Eisenhower took office in 1953, Strauss became chairman of the AEC. From this influential post Strauss plotted to eliminate Oppenheimer's influence over atomic energy policy. The campaign culminated in a hearing over whether to renew Oppenheimer's security clearance. The proceeding was a sham. Strauss abused his power to stack the panel and manage its operation. Oppenheimer was stripped of his security clearance, and his role in government was over. However, Strauss was unable to remove Oppenheimer from the directorship of the Institute. Oppenheimer remained in office until his retirement in 1966, one year prior to his death.[58]

During Oppenheimer's long Institute tenure there was naturally a large turnover in the faculty. Early in its existence the School of Historical Studies moved to sever its relations with Mitrany. When Earle died in 1954, no trace remained from the original school set up by Flexner to pursue an interdisciplinary attack on economics. At this time only Panofsky and Meritt were still in place from the first generation of Humanistic Studies.[59]

Most of the Oppenheimer appointments to Historical Studies were historians of various specialities. The school's best-known professor was George Kennan who had had a distinguished career in foreign service that included an influential 1947 essay on Soviet containment. Kennan was one of the few diplomats in agreement with Oppenheimer's position on the hydrogen bomb. By the time Kennan fell out of favor with the Truman administration in 1950, he and Oppenheimer were mutual admirers. Kennan then accepted an Institute membership that he left in 1952 for an abortive stint as United States ambassador to the Soviet Union. In 1955 Oppenheimer proposed Kennan for a professorship in Historical Studies.[60]

Kennan was an unusual candidate. He lacked a PhD and his writings were of a different nature than those of university professors. From their extensive discussions, Oppenheimer was convinced that Kennan would excel at scholarly work. Mathematicians, such as von Neumann and Montgomery, argued that Kennan was unqualified to be an Institute professor. Kennan joined the faculty in 1956. The following year he received both a Pulitzer Prize and a National Book Award.[61]

The Oppenheimer years (1947–1966) were a period of considerable transition for the School of Mathematics. Veblen became emeritus in 1950. Einstein and Weyl died in 1955. That same year von Neumann was diagnosed with cancer. He lived another one and a half years, and, during this period, accepted a position at the University of California. By then just Morse remained from the first generation of the school's professors. In the 1950s the number of professors who were mathematicians increased to eight. Joining Morse on the faculty were Montgomery, Selberg, Gödel, Hassler Whitney, Arne Beurling, Armand Borel, and André Weil. Meanwhile Oppenheimer added physicists Pais, Freeman Dyson, Chen Ning Yang, Bengt Stromgren, and T. D. Lee. The two groups, while remaining in the same school, functioned separately.[62]

With Morse slated to retire in 1962, the mathematicians met to discuss various candidates. Armand Borel offers his recollection of the meeting:

> I remember distinctly myself describing in particular the work of Grothendieck, Milnor and Harish-Chandra, and even being thanked by Oppenheimer for all that information. As it became clear that our top candidates were Milnor and Harish-Chandra and we were nearing a vote, then Oppenheimer thought it proper to leave the room and his last remark, already on the door step, was that a nomination of Harish-Chandra would go smoothly, but one of Milnor might cause problems (he was not more specific, I am sure of that).[63]

John Milnor was a 31-year-old topologist who had obtained striking results about seven-dimensional spheres. His consideration generated controversy because Milnor was on the faculty of Princeton University.

The mathematicians decided to go forward with Milnor. When the entire faculty met, it became evident that the disciplines had different understandings of Institute tradition and recruiting regulations. Physicists and historical studies professors regarded university faculty as off-limits. Mathematicians knew of no prohibition. Given the absence of any available written record, the confusion was natural. Opinions were heated, and the long session was charged with acrimony.* After the faculty meeting Leidesdorf weighed in with his letter invoking the pledge that "we would never interfere with the Princeton faculty." (See Chapter 6 for full text and analysis). The Institute mathematicians, unaware of the 1932 action by the University Curriculum Committee, initiated their own discussions with colleagues in the Princeton mathematics department. The prevailing view was that a free market should exist. Meanwhile, the university president phoned Oppenheimer to request that no offer be made to Milnor.[64]

The Institute faculty gathered again to discuss the Milnor nomination and Leidesdorf's letter. As with the previous meeting there was considerable rancor. Art historian Millard Meiss attempted to bring clarity to the extended debate. Under his successful motion, university faculty would henceforth be eligible for Institute positions, but such moves were to be rare and taken with great sensitivity. Oppenheimer agreed to bring the Meiss proposal to the trustees, but decided not to forward the nomination of Milnor. It is understandable that the trustees acted with considerable deference to the view of Leidesdorf. He was the chairman and the only remaining member who had served on the Board since its creation. Who was to argue with his self-assured rendition of history? The trustees *reaffirmed* the policy of noninterference with the university. The Milnor nomination was never moved.[65]

Review of these events may not convey the enmity that arose out of the interactions. The mathematicians were furious. Borel explained,

> It was not just the decision of the trustees, but the way the matter had been handled and the breakdown in relations within the faculty . . . , the ruling from on high by the board, without bothering to have a meaningful discussion with us, bluntly disregarding our wishes, as well as those of the faculty as expressed by the Meiss motion, all this chiefly on the basis of a rather flimsy recollection of the chairman of the board, promoted to the status of an irrevocable pledge.[66]

The mathematicians were particularly incensed at Oppenheimer who had obstructed them at every stage.

---

*Minutes of this meeting are unavailable. While the date is beyond the IAS thirty-year limit, records remain closed on personnel matters and letters of recommendation as long as the subject or author survives.

It was André Weil who devised the therapy to get beyond the Milnor trauma. Rather than propose a single substitute, the mathematicians nominated both Harish-Chandra and Lars Hormander. When the appointments succeeded, the mathematicians did feel better, but the school was in a terrible state. Some of the mathematicians and physicists were no longer on speaking terms. A partial resolution took place in 1965 when the physicists split off to form the School of Natural Sciences. By this time Lee and Pais had left the Institute, and Yang was on his way out.[67]

Bitter feelings persisted between the director and the mathematicians. These irreconcilable differences were among the reasons behind Oppenheimer's slightly premature retirement. Over Oppenheimer's 19 years he adhered to Flexner's fundamental design for the Institute. Some changes did occur during this period. One school bifurcated and the two others merged. Faculty salaries were equalized, first at $18,000. Significant infrastructure improvements included a new library, additional office space, and a housing project for members. After the Milnor struggle a formal policy was adopted making it possible, in theory, to recruit faculty from the university.[68]

Prior to selecting Oppenheimer's successor, the trustees, in 1965, initiated a critical review of the Institute. Evaluations and suggestions were solicited from a number of distinguished outside scholars. Everything was on the table, even consideration of a merger into the university. One outcome of this process was a decision that the curriculum of mathematics, physics, and historical studies should be broadened. The fourth director, Harvard economist Carl Kaysen, arrived with a mandate to create a school in social science. The faculty had no voice in these decisions.[69]

In many respects Kaysen's background was similar to that of Oppenheimer at the time of his appointment. Both were in their mid-forties, held prestigious university positions, and had served in influential government posts. Kaysen was among the so-called "best and brightest" who, in 1961, were called to the Kennedy Administration in Washington. After his two years of operating in the rarefied atmosphere of the National Security Council, Kaysen was unlikely to offer much deference in his dealings with Institute professors.[70]

When Kaysen took over in 1966, only epigrapher Ben Meritt remained from the Flexner days. The faculty numbered nine in historical studies, nine in mathematics, and four in natural sciences. Dissatisfaction with Kaysen arose independently within the Schools of Mathematics and Historical Studies. The grievances involved differences over budget priorities and personnel.

Tension increased as Kaysen embarked on establishing the School of Social Science. It erupted in 1973, leaking onto the front page of the *New York Times*.[71] One year later an article by Landon Jones appeared in *The Atlantic* entitled "Bad Days on Mount Olympus." These accounts provide more details on the conflict than are contained in the following summary. A retrospective insider's view can be found in Morton White's *A Philosopher's Story*.

Kaysen's first social science nomination went smoothly. Anthropologist Clifford Geertz joined the faculty in 1970. The following year Kaysen unsuccessfully floated the name of a psychologist. Then, in 1972, Kaysen and Geertz proposed Berkeley sociologist Robert Bellah. Professors in historical studies and mathematics decided that Bellah fell short of Institute standards. After some heated campaigning eight faculty voted in favor of the appointment and 13 opposed, with three abstentions. Kaysen faced a major setback for his social science plan. Under Aydelotte and Oppenheimer, names had moved forward only after receiving the endorsement of the faculty. Yet these were the practices of individual directors rather than the result of statutory obligations. Kaysen elected to defy the faculty and take the nomination of Bellah to the trustees.

The Board sided with its director, approving an offer to Bellah. Members of the faculty were outraged at both Kaysen and the trustees. Fourteen professors joined together in petitioning for an outside review of the director. It was at this point that the story broke in the *New York Times*. The Board responded with some unsuccessful attempts to deflect the confrontation.

The mutual disrespect among the faculty, director, and trustees was evident from contemporaneous comments. Mathematics professor Armand Borel, speaking for an expanded group of dissidents to a committee of trustees, said:

> It is a fact that at least 17 members of the faculty of 27 have lost confidence in the director, to the point of finding it futile to transact any business with him. It is our considered judgment that we cannot trust his moral integrity, administrative capacity and intellectual judgment, and, therefore, that it is pointless to engage in serious discussions or negotiations with him on matters pertaining to the Institute at large.[72]

Trustee J. Richardson Dilworth alleged that Kaysen's detractors "regard[ed] the director as a sort of janitor who cleans the buildings while they run the place."[73] Mathematician André Weil mocked the trustees' financial support for the Institute. As for Bellah, he first accepted the offer and later withdrew.

A dispute over professional qualifications had escalated into the Institute's greatest crisis. Such occurrences are not uncommon in academic settings. The "paradise of scholars" carried no immunity. Personal animosity pervaded relationships among the small Institute community. It would take many years to recover. One constructive outcome was the adoption of rules of governance, setting forth the roles of the faculty, director, and trustees in the hiring process. Kaysen soldiered on for two difficult years before announcing his resignation. He did succeed in making an additional appointment in social science. Harvard professor Albert Hirschman brought strength to the Institute in its earlier area of economics and politics.

For its next director the Board selected a more seasoned academic administrator. Harry Woolf, an historian of science, was the provost at Johns Hopkins University. Woolf was in his mid-fifties, about ten years older than his two predecessors. He took over an Institute beset by dissension and financial difficulties, serving as director from 1976 to 1987. Mathematics professor emeritus Atle Selberg recalls this period as the most pleasant and peaceful of his nearly sixty years at the Institute.[74]

In 1986 investment banker James Wolfensohn succeeded Dilworth as chairman of the Board. One year later physicist Marvin Goldberger left the presidency of Caltech to become the Institute's fifth director. Goldberger had been a member in the early days of the School of Natural Sciences. He served as director for just four years, retiring at the age of 68. To replace Goldberger the Institute selected mathematician Phillip Griffiths. Griffiths was well acquainted with Princeton. He had obtained his PhD from the university and then gone on to a distinguished research career with memberships at the Institute and faculty positions at Berkeley, Princeton, and Harvard. Griffiths left Harvard in 1983 to became provost of Duke. He moved comfortably into his first administrative post, but regretted that the new responsibilities left little time for scholarly work. When Goldberger was slated to retire, School of Mathematics faculty sounded out Griffiths about becoming director. The opportunity to continue his scientific life was appealing. Wolfensohn followed up with a recruiting trip to North Carolina. Griffiths became director in 1991 at the age of 52.[75]

Griffiths worked effectively with Wolfensohn and the Board to advance the Institute. A number of trustees made substantial donations, increasing endowment and facilitating major infrastructure upgrades. Most significant were the measured steps taken toward the Institute's first curricular expansion since the creation of the School of Social Science. In the intervening years there had been exciting progress in subjects, such as theoretical biology and computer science, that were outside the expertise of current faculty.

To staff a new school, Institute regulations required a minimum of three professors. One lesson from the Kaysen debacle was the wisdom of proceeding more incrementally. Rather than committing to any new schools, Griffiths expanded the scope of existing programs. Homes for a new professor in the fields of computer science, biology, and East Asian studies were established in the Schools of Mathematics, Natural Sciences, and Historical Studies respectively. This approach broadened the Institute's coverage while leaving open the possibility of creating future schools on already existing foundations.

In 2004 Griffiths left his administrative post and joined the School of Mathematics faculty. Physicist Peter Goddard took over as director. Endowment stood at approximately $475,000,000. The annual operating budget was $41,000,000 with over half for salaries and another $6,000,000 to stipends. With no income from tuition and fees, the $13,000,000 in donations and grants stood as an important line. The Institute has four schools, 26 professors, and about 190 members. As for the future, Goddard has no plans for major changes. He is strongly committed to the original mission and conception, stressing the importance of "curiosity-driven research."[76]

It is a tribute to Flexner that his seven successors have been so faithful to his vision. Every director has striven to provide an ideal environment for high level research by a small number of eminent permanent professors and temporary members. The only modifications occurred in Flexner's early years when the profile for the temporary people underwent an evolution. Originally, in 1930, the Institute planned to admit graduate students who were to work toward the doctoral degree. In 1932 Flexner upgraded the admission requirement to a PhD. At this point Flexner expected to target new PhDs who would come to the Institute to pursue research under the supervision of individual faculty. When Veblen was authorized to begin admitting "students," he took the liberty of regarding the PhD as a minimum standard, bringing in scholars at varying career stages. Flexner was so impressed by the harmony of this first group that he embraced Veblen's modification, which remains in effect today.

It should be noted that some variation has existed among the schools in the connection of members to faculty. The School of Mathematics has always followed Veblen in regarding members as independent scholars with freedom to pursue whatever program they choose. In other schools the work of members has been more closely linked with that of individual faculty. The remainder of this chapter will deal primarily with the School of Mathematics.

Flexner and Veblen shared a strong belief in the notion of a community of scholars. In the School of Mathematics the Institute provides little formal structure. Its role is to bring together capable scholars in a favorable setting. The expectation is that under these circumstances research activity is stimulated and interaction thrives. Each year about sixty members are invited. Some seminars are organized by faculty while others seem to generate spontaneously.

When Flexner conceived the Institute, his objective was to elevate research in America. The members were a key ingredient of the program. Under Flexner's design the Institute was to prepare young scholars so that the rising generation would be stronger. The vast advances of mathematics in America, and worldwide, during the past 75 years have been stunning. It is beyond the scope of this project to evaluate the contribution of the Institute to this success. However, a few anecdotal cases illustrate the type of influence it has exerted.

Raoul Bott became a member in 1949. He had just completed a thesis on electrical networks at Carnegie Institute of Technology. During his two years at the Institute Bott essentially retooled, learning topology and Morse Theory. His subsequent work in these areas was recognized by a National Medal of Science in 1987.[77]

Two of the greatest mathematical results of the twentieth century were obtained by young topologists shortly after completing memberships. Stephen Smale's 1960 proof of the higher dimensional Poincaré Conjecture came just a few months following a one and a half year stay at the Institute. The four-dimensional version of this theorem was completed by Michael Freedman in 1982, one year after his residence. Both Smale and Freedman received the Fields Medal, the most prestigious award in mathematics. Flexner would have loved these chronologies.

Not every membership is as decisive as Bott's or as timely as those of Smale and Freedman. With sixty invitations at their disposal, the faculty is in a position to take risks, and they do. In some cases the selection is based primarily on potential. It would be unreasonable to expect a high percentage of miracles. However, each year dozens of mathematicians are permitted to conduct research under ideal conditions. Prior to the existence of the Institute, such opportunities did not exist. Many other operations have arisen since, inspired, in part, by the success at Princeton.[78]

While the Flexner-Veblen conception of the postdoctoral member is widely applauded, there has been some criticism of the design for the faculty component. The basic premise of these arguments is that the Institute's conditions are too ideal to sustain a lifetime of scholarly creation. Physicist Richard Feynman gave his perspective:

> When I was at Princeton in the 1940s I could see what happened to those great minds at the Institute for Advanced Study, who had been specially selected for their tremendous brains and were now given this opportunity to sit in this lovely house by the woods there, with no classes to teach, with no obligations whatsoever. These poor bastards could now sit and think clearly all by themselves, OK? So they don't get an idea for a while: They have every opportunity to do something, and they're not getting any ideas. I believe that in a situation like this a kind of guilt or depression worms inside of you, and you begin to *worry* about not getting any ideas. And nothing happens. Still no ideas come.
>
> Nothing happens because there's not enough *real* activity and challenge: You're not in contact with the experimental guys. You don't have to answer questions from the students. Nothing!
>
> In any thinking process there are moments when everything is going good and you've got wonderful ideas. Teaching is an interruption, and so it's the greatest pain in the world. And then there are the *longer* periods of time when not much is coming to you. You're not getting any ideas, and if you're doing nothing at all, it drives you nuts! You can't even say "I'm teaching my class."[79]

Feynman declined an Institute faculty position, preferring the environment of a university. Other outstanding scholars have viewed the Institute's conditions as well suited to their work. The existence of these contrary opinions may indicate that there is no single set of circumstances that best serves every powerful creative mind. It is impossible to perform the experiment of, say, switching Lefschetz and Alexander between their university and Institute positions in 1933, and then comparing the resulting bodies of work.

An argument could be made that, with the exception of von Neumann, all of the original mathematics faculty did their best work prior to their appointment at the Institute. However, von Neumann was considerably younger than his colleagues, who each arrived in their mid-forties or fifties. Whether or not mathematicians do their best work earlier in life is itself a controversial question. Recently deceased Institute astrophysicist John Bahcall offered an interesting perspective on the difficulty of performing an accounting on scholarship at this level:

> Everyone who gets to the Institute as a permanent professor gets here because he's done—in all cases—two important things. Otherwise you don't

get here. It's very hard to do more than two important things in the sciences. Or even to do one. So often your best work is done before you get here permanently.[80]

Even if they never obtain that third, or fourth, great result, the true value of the faculty may lie in their everyday role as an extraordinary resource for the members. Young scholars have ready, if daunting, access to several of the world's greatest minds. There may even be some inspiration derived from walking in the footsteps of Einstein and Gödel. Without the faculty, the Institute would be a very different place.

Under Flexner's original analysis, the faculty was the key to the success of the Institute. He had to get "men and women of genius, of unusual talent, and of high devotion."[81] Through the influence of Veblen, the School of Mathematics began with the legendary group of Einstein, Veblen, Alexander, von Neumann, Weyl, and Morse.

Following Flexner's retirement, the faculty became largely self-perpetuating, subject to the oversight of the director and trustees. The history of the Institute reveals an overarching determination to bring in the world's best scholars. The faculty's resolve in this regard overcame Flexner's noncitizen quota to embrace the founding ideals of Bamberger and Fuld. Two striking features of the list of past mathematics professors are the international composition and the consistent level of quality. The second generation of Montgomery, Selberg, Gödel, Whitney, Beurling, Borel, and Weil made seminal contributions to the development of mathematics. Currently the Institute faculty includes six Fields Medalists: Bombieri, Deligne, Bourgain, Voevodsky, Selberg (emeritus), and Witten (School of Natural Sciences). In 1993 the academic governance procedures were modified. These changes, establishing mechanisms for the consideration of outside and opposing views, should help ensure that future faculty appointments maintain the existing tradition of distinction.[82]

SOURCES AND ACKNOWLEDGMENTS

In 1955 Beatrice Stern was commissioned by Robert Oppenheimer to commence research on the history of the Institute for Advanced Study. Stern assembled the available records and interviewed the surviving players. Her 715-page manuscript documented, in considerable detail, the Institute's prehistory and its first twenty years. Stern's forthright account occasionally portrayed some faculty and administrators in an unfavorable light. The Institute authorities reacted by suppressing her manuscript. For many years neither the narrative, nor the records on which it was based, were available to outside scholars. Most accounts of the Institute's history depended heavily on the few copies of Stern's manuscript that had reached libraries and a microfilm version that was included among Oppenheimer's papers at the Library of Congress.

During the 1980s the Institute began gathering its materials into the Archives of the Institute for Advanced Study. By 1999, when I began my research, restrictions on access to the Stern manuscript were relaxed. At this time my focus was on the amazing first generation of faculty in the School of Mathematics. I was especially interested in Flexner's decision to begin with mathematics. I met with the Director, Phillip Griffiths, to discuss my vague plans for a book about the development of mathematics at the Institute. We agreed that my book should neither be construed as authorized or unauthorized. The Institute's staff has graciously cooperated in opening their records, but my work has been independent and outside their review.

One valuable resource was an earlier article by Armand Borel on the history of the School of Mathematics. As an emeritus professor of mathematics, he had access to the archive. I was fortunate to have several discussions with Borel prior to his death in 2003. He suggested that, along with mathematics, I include the emergence of the other schools. The result is a book about the Institute's origin and first ten years, with emphasis on the School of Mathematics and the faculty.

My primary sources are the contemporaneous documents stored in the Archives of the Institute for Advanced Study, the Library of Congress, and other repositories. These records include most of the materials that were available to Stern, as well as documents that were not. The conclusions reached by Stern and myself are compatible on many major aspects of the Institute's development. Two notable exceptions are the creation of the postdoctoral policy and the portrayal of Oswald Veblen.

A project of this nature owes much to the archival staff who locate the appropriate materials. I would like to thank Fred Bauman (Library of Congress), Rob Cox (American Philosophical Library), Chris Densmore (Friends Historical Library), and Patrice Donoghue (Harvard University Archives). Over the past six years I have been helped, time and time again, by Erica Mosner, Marcia Tucker, Momota Ganguli, and especially Lisa Coats at the Institute for Advanced Study.

Photographs are vital to documenting history. In depicting members of the Institute faculty, I have sought images that were created at about the time the subject came under Flexner's consideration. The following people were instrumental in locating and making available the photographs in this book: Jeffrey Moy (The Newark Museum), Lesley Schoenfeld (Harvard Law School Library), Shari Kenfield (Princeton University Department of Art and Archaeology), James Stimpert (Johns Hopkins University Archivist), Carl Ashley (American Historical Association), Christine Thivierge and Ina Lindemann (American Mathematical Society), Renee Mastrocco (Rockefeller University Archives), Marcia Tucker (Institute for Advanced Study), Nina Weyl, Ellen Viner Seiler, and Marina von Neumann Whitman.

A number of individuals were of invaluable assistance in a variety of ways. Tilmann Glimm and Margrit Nash translated letters from German to English. Bonna Wescoat, Herb Benario, and Eric Varner patiently responded to my questions about classical studies scholarship. R. Narasimhan connected me to important sources.

I would especially like to thank several people who kindly sat for interviews: Armand Borel, Peter Goddard, Phillip Griffiths, Louise Morse, Atle Selberg, and Morton White. Emory University supported a research leave that enabled me to launch the project.

Finally, I am most grateful to Albert Lewis, Ellen Neidle, and Paul Sally for providing encouragement and suggesting improvements in my exposition.

ARCHIVES

[ABP] Armand Borel's personal papers

[APS] Simon Flexner Papers, American Philosophical Society

[FHL] Frank Aydelotte Papers, Friends Historical Library of Swarthmore College

[HUA] G. D. Birkhoff Papers, Harvard University Archive

[IAS] Archives of the Institute for Advanced Study

[LCF] Abraham Flexner Papers, Manuscript Division, Library of Congress

[LCN] John von Neumann Papers, Manuscript Division, Library of Congress

[LCV] Oswald Veblen Papers, Manuscript Division, Library of Congress

[SGM] Seeley G. Mudd Manuscript Library, Princeton University

ENDNOTES

For most endnotes on nonquoted material, superscripts appear at the end of paragraphs. Archival sources are prefaced by three upper case letters in square brackets using the code above. Notes for books, articles, and other non-archival sources are introduced by the author's name and a partial title. Full bibliographic information for these non-archival materials is available in the list following the endnotes.

PREFACE

[1] [IAS] 1/24/38 Trustees minutes
[2] [IAS] 4/13/36 Trustees minutes
[3] [LCV] 2/17/32 Weyl to Veblen, translation by Tilmann Glimm

CHAPTER ONE

[1] References on Abraham Flexner: Bonner, *Iconoclast*; Flexner, *Autobiography*; Flexner, *American Saga*
[2] [LCF] 6/22/31 Abe to Eleanor and Jeannie
[3] References on Gilman and Johns Hopkins: Flexner, *Daniel Coit Gilman*; Franklin, *Life*; French, *History of the University*; Hawkins, *Pioneer*
[4] Flexner, *Daniel Coit Gilman*, p. 52
[5] Birkhoff, *Mathematics at Harvard*; Richardson, "PhD"
[6] Hawkins, *Pioneer*, p. 67
[7] Parshall and Rowe, *Emergence*, p. 58

[8] Reference on Sylvester: Parshall and Rowe, *Emergence*

[9] Flexner, *Autobiography*, p. 33

[10] Reference on Simon Flexner: Flexner, *American Saga*

[11] Hawkins, *Pioneer*, p. 148

[12] References on Klein and his recruitment by Hopkins: Parshall and Rowe, *Emergence*; Reid, "Road Not Taken"

[13] Parshall and Rowe, *Emergence*, p. 187

[14] Reference on Klein's influence: Parshall and Rowe, *Emergence*

[15] Franklin, *Life*, p. 389

[16] Flexner, *American Saga*, p. 379

CHAPTER TWO

[1] Reference on John D. Rockefeller: Chernow, *Titan*

[2] References on the University of Chicago: Goodspeed, *History*; Storr, *Harper's University*

[3] References on E. H. Moore: Parshall and Rowe, *Emergence*; Archibald, *A Semicentennial History*, pp. 144–150

[4] Parshall and Rowe, *Emergence*, p. 284

[5] Parshall and Rowe, *Emergence*, p. 286

[6] Archibald, *A Semicentennial History*, p. 108

[7] Richardson, "PhD"

[8] Reference on Gates: Chernow, *Titan*

[9] Chernow, *Titan*, p. 468

[10] Chernow, *Titan*, p. 470

[11] References on the Rockefeller Institute for Medical Research: Corner, *History of Rockefeller*; Fosdick, *Story of the Rockefeller*; Chernow, *Titan*

[12] quoted in Chernow, *Titan*, p. 479, from the *Saint-Louis Post Dispatch*, 7/8/36

[13] References on the General Education Board: Bonner, *Iconoclast*; Chernow, *Titan*; Flexner, *Autobiography*; Fosdick, *Story of the Rockefeller*. References on the General Education Board and racism: DuBois, *Autobiography*; Link, *Paradox*

[14] Flexner, *Medical Education*

[15] Fosdick, *Story of the Rockefeller*, p. 100

[16] References on Flexner-Lowell enmity: Bonner, *Iconoclast*; [APS] 4/21/32 Abe to Simon

[17] Fosdick, *Story of the Rockefeller*, pp. 98–99

[18] References on Rose: Fosdick, *Story of the Rockefeller*; Siegmund-Schultze, *Rockefeller*

[19] Bonner, *Iconoclast*, pp. 210–211; Porter, *From Intellectual Sanctuary*, pp. 55–57

CHAPTER THREE

[1] References on the Bambergers: [LCF] 11/8/44 Lavinia Bamberger to Flexner; Bennett, "Do the Wise Thing"; [LCF] 8-page recollection apparently by Hardin; [LCF] 3-page obituary; Stern, *History*

[2] Bennett, "Do the Wise Thing," p. 72

[3] *New York Times* 6/30/29, 8/22/29

[4] References on Bamberger-Fuld endowment: [IAS] 1955 Maass, The Founding and Early History of the Institute; Flexner, *Autobiography*; Stern, *History*

[5]Flexner, *Autobiography*, p. 180

[6][LCF] 11/28/44 Maass to Flexner

[7][IAS] 1/20/30 Memorandum

[8][IAS] 2/24/30 Flexner to Maass; [IAS] 2/24/30 Flexner to Leidesdorf; [IAS] 3/8/30 Flexner to Bamberger

[9]Kennedy, *Freedom from Fear*, p. 58

[10]Flexner, *Universities*, pp. 162, 160

[11]Flexner, *Universities*, pp. 206–207

[12]Flexner, "A Modern University"

[13]Flexner, *Universities*, pp. 213–214

[14] [IAS] 4/23/30 Flexner draft for letter to Trustees; [IAS] 5/5/30 Bamberger to Flexner; [IAS] 5/9/30 Flexner to Bamberger

[15][IAS] 4/23/30 Flexner draft for letter to Trustees

[16][IAS] 6/4/30

[17][APS] 2/31 Simon to Abe

[18][IAS] undated slate of nominees for Board

[19]Reference on Aydelotte: Blanshard, *Frank Aydelotte*

[20][APS] 6/17/30 Abe to Simon

[21][APS] 7/10/30 Abe to Simon

[22]Flexner, *Autobiography*, p. 136

[23][IAS] 9/17/30 Flexner draft of by-laws; [IAS] 10/7/30 draft of by-laws

[24][IAS] 10/10/30 Trustees minutes

[25][IAS] 10/10/30 edited Director's report

[26][IAS] 10/10/30 edited Director's report

[27][APS] 2/31 Simon to Abe

[28][APS] 3/17/31 Abe to Simon

CHAPTER FOUR

[1][APS] 12/8/30 Abe to Simon; [IAS] 12/11/30 Flexner to Aydelotte; [IAS] 12/16/30 Aydelotte to Flexner; [IAS] 12/23/30 Weed to Flexner; [IAS] 1/12/31 Sabin to Flexner

[2][IAS] 12/1/30 Flexner memo of interview with Smith

[3][IAS] 1/16/31 Trustees minutes

[4][IAS] 2/11/31 Flexner to Bamberger and Fuld

[5][IAS] 2/16/31 Memo on Jameson advice; [IAS] 3/3/31 Jameson to Flexner; [IAS] 4/25/31 Jameson to Flexner; [IAS] 6/28/31 Beard to Flexner; [IAS] 7/18/31 Flexner to Beard; [IAS] 7/22/31 Beard to Flexner; [IAS] 7/27/31 Flexner to Beard; [IAS] 8/3/31 Beard to Flexner; Reference on Beard: Barrow, *More than a Historian*

[6]Reference on Warburg: Chernow, *The Warburgs*

[7][LCF] 2/7/31 Abe to Eleanor; [IAS] 2/11/31 Flexner to Bamberger and Fuld; [APS] 2/11/31 Abe to Simon

[8][IAS] 3/1/31 Armstrong to Flexner

[9][IAS] 2/16/31 Morey to Flexner

[10][IAS] 2/18/31 Lefschetz to Flexner

[11]Siegmund-Schultze, *Rockefeller*

[12][IAS] 2/18/31 Young to Flexner

[13] [IAS] 2/24/31 Flexner to Bamberger

[14] [IAS] 3/17/31 G. D. Birkhoff to Flexner

[15] References on G. D. Birkhoff: Veblen, *George David Birkhoff*; Archibald, *A Semicentennial History*, pp. 212–218; Birkhoff, *Some Leaders*; Birkhoff, *Mathematics at Harvard*

[16] Aspray, *Emergence*, p. 197

[17] [HUA] 1/19/09 Fine to Birkhoff; [HUA] 1/25/09 Bliss to Birkhoff

[18] Veblen, *George David Birkhoff*

[19] [HUA] 2/1/11 E. H. Moore to Birkhoff

[20] [HUA] 4/8/12 Bôcher to Birkhoff; [HUA] 4/20/12 Fine to Birkhoff

[21] Diacu and Holmes, *Celestial Encounters*

[22] Reid, *Courant*, pp. 45–46

[23] References on Viner: Shils, *Remembering*, pp. 533–547; Breit and Hirsch, *Lives of the Laureates*, pp. 70, 81–82

[24] [APS] 3/17/31 Abe to Simon

[25] [LCF] 3/29/31 Abe to Jean; Albers, *International*

[26] [LCF] 4/14/31 Abe to wife Anne

[27] [LCF] 4/18/31 Abe to daughter Jean

[28] [LCF] 4/19/31 Abe to wife Anne

[29] [LCF] 5/7/31 Abe to Anne; [LCF] 5/12/31 Abe to Jean; [IAS] 5/31/31 Schumpeter to Flexner

[30] [IAS] 6/10/24 Veblen to Flexner

[31] [LCV] undated Fine, Eisenhart, and Veblen to Hibben; [LCV] Institute for Mathematical Research at Princeton; [LCV] 10/21/25 Veblen to Thorkelson

[32] Aspray, *Emergence*; Porter, *From Intellectual Sanctuary*

[33] [SMM] The Princeton Mathematics Community in the 1930s: An Oral History Project, on the web at http://infoshare1.princeton.edu/libraries/firestone/rbsc/finding_aids/mathoral/math.html

[34] [IAS] 1/21/30 Flexner to Veblen; [IAS] 1/24/30 Veblen to Flexner; [IAS] 1/27/30 Flexner to Veblen; [IAS] 1/29/30 Veblen to Flexner

[35] [IAS] 6/10/30 Veblen to Flexner; [IAS] 6/19/31 Veblen to Flexner

[36] [FHL] 6/1/31 Aydelotte to Flexner; [FHL] 6/2/31 Flexner to Aydelotte; [IAS] 6/8/31 Flexner to Maass

[37] [IAS] 6/9/31 Maass to Flexner; [IAS] 6/10/31 Flexner to Maass; [IAS] 6/15/31 Maass to Flexner; [IAS] 6/16/31 Flexner to Maass

[38] [LCF] 6/22/31 Abe to Eleanor and Jeannie

[39] [IAS] 9/26/31 Confidential (report) to the Trustees

[40] Flexner, *Autobiography*, p. 235

[41] [IAS] 9/22/31 Flexner to Frankfurter

[42] [IAS] 6/19/31 Veblen to Flexner; [IAS] 9/21/31 Frankfurter to Flexner; [IAS] 10/22/31 Flexner to Straus

[43] [IAS] 11/5/31 Birkhoff to Flexner

[44] [LCF] 11/21/31 Abe to Jean; [IAS] 1/7/32 Frankfurter to Flexner; [IAS] 1/11/32 Flexner to Frankfurter

[45] [IAS] 11/5/31 Flexner to Bamberger; [IAS] 11/6/31 Bamberger to Flexner

[46] [IAS] 12/7/31 Committee on Site minutes; [IAS] mailing list for letters to Birkhoff and others; [IAS] 12/17/31 Flexner to Birkhoff

[47] [IAS] 12/9/31 Flexner to Veblen

[48] [LCV] 10/16/31 Weyl to Veblen; [LCV] 10/28/31 Veblen to Weyl

[49] [IAS] 2/20/32 Flexner to Maass

[50] [IAS] 12/15/31 Veblen to Flexner

[51] [IAS] 12/16/31 Flexner to Veblen

[52] [IAS] 1/1/32 Veblen to Flexner

[53] [IAS] 12/17/31 Maass to Hardin; [IAS] 1/11/32 Docket for Trustees meeting; [IAS] 1/11/32 Trustees minutes

[54] [IAS] 8/15/31 Beard to Flexner; [IAS] 8/30/31 Aydelotte to Flexner; [IAS] 9/21/31 Frankfurter to Flexner

[55] [LCF] 11/21/31 Abe to Jean

CHAPTER FIVE

[1] References on Einstein: Clark, *Einstein: The Life and Times*; Pais, *'Subtle is the Lord'*; Pais, *Einstein Lived Here*; Brian, *Einstein: A Life*; Frank, *Einstein: His Life*

[2] Flexner, *Autobiography*, p. 257

[3] Clark, *Einstein*, p. 232

[4] [IAS] 12/8/30 Flexner to Bamberger

[5] Reference on Hitler and Nazi Party: Kershaw, *Hitler*

[6] *New York Times* 12/12/30, p. 1

[7] [IAS] 7/25/32 Millikan to Flexner

[8] Flexner, *Autobiography*, pp. 250–251

[9] [LCF] 2/3/32 Abe to Jean; [APS] 2/8/32 Simon to Abe

[10] [IAS] 2/13/32 Flexner to Bamberger

[11] [IAS] 7/25/32 Millikan to Flexner

[12] [LCF] 2/3/32 Abe to Jean; [IAS] 8/20/32 Flexner memorandum regarding Weyl

[13] References on Weyl: Sigurdsson, *Hermann Weyl*; Newman, "Hermann Weyl"; Reid, *Courant*; Reid, *Hilbert*

[14] Weyl, "David Hilbert," p. 614

[15] Sigurdsson, *Hermann Weyl*, pp. 61–62, quoted from Weyl's "Lecture at the Bicentennial Conference"

[16] Hawkins, *Emergence*

[17] [LCV] 3/22/28 Veblen to Weyl; [LCV] 11/28/28 Fine to Veblen

[18] [LCF] 2/3/32 Flexner to Mrs. Bailey

[19] Reference on economic situation: Kennedy, *Freedom from Fear*

[20] [IAS] 2/20/32 Bamberger to Flexner

[21] [IAS] 2/13/32 Flexner to Bamberger

[22] [IAS] 2/12/32 Flexner to Birkhoff

[23] [IAS] 2/15/32 Flexner to Weyl

[24] [IAS] 2/15/32 Birkhoff to Flexner; [IAS] 2/21/32 Birkhoff to Flexner; [IAS] 2/23/32 Flexner to Birkhoff

[25] [IAS] 2/20/32 Flexner to Maass; [IAS] 2/25/32 Flexner to Birkhoff

[26] [IAS] 3/1/32 Flexner to Bamberger

[27] [IAS] 2/29/32 Flexner to Birkhoff

[28] [IAS] 3/4/32 Birkhoff to Flexner; [HUA] 3/7?/32 Mrs. Birkhoff to Coolidge

[29] [HUA] 3/8/32 Coolidge to Birkhoff; [HUA] 3/9/32 Coolidge to Mrs. Birkhoff; References on Flexner-Lowell enmity: Bonner, *Iconoclast*; [APS] 4/21/32 Abe to Simon

[30] [HUA] 3/9/32 Richardson to Birkhoff

[31] [IAS] 3/7/32 Birkhoff to Flexner; [HUA] 3/9/32 Oliver to Birkhoff

[32] [HUA] 3/8/32 Coolidge to Birkhoff

[33] [IAS] 3/9/32 Birkhoff to Flexner; [IAS] 3/9/32 Flexner to Birkhoff; [IAS] 3/10/32 Flexner to Birkhoff

[34] [IAS] 3/10/32 Birkhoff to Flexner; [IAS] 3/11/32 Flexner to Birkhoff

[35] [IAS] 3/20/32 Birkhoff to Flexner; [IAS] 3/21/32 Flexner to Birkhoff; [IAS] 3/23/32 Flexner to Mrs. Fuld

[36] [IAS] 3/28/32 Birkhoff to Flexner

[37] [LCV] 4/2/32 Birkhoff to Veblen

[38] [HUA] 5/9/33 Lowell to Birkhoff; [HUA] 5/20/33 Birkhoff appointment as Perkins Professor

[39] Birkhoff, *Mathematics at Harvard*

[40] [APS] 4/20/32? Simon to Abe

[41] [APS] 4/21/32 Abe to Simon

[42] [HUA] 3/5/32 Veblen to Birkhoff; [LCV] 4/2/32 Birkhoff to Veblen; [HUA] 4/24/32 Veblen to Birkhoff

[43] [LCV] 1/6/32 Weyl to Veblen

[44] [LCV] 2/17/32 Weyl to Veblen, translation by Tilmann Glimm

[45] [IAS] 3/5/32 Weyl to Flexner

[46] [LCV] 3/5/32 Veblen to Weyl

[47] References on Veblen: Montgomery, *Oswald Veblen*; Archibald, *A Semicentennial History*, pp. 206–211; Aspray, *Emergence*

[48] Veblen, "Theory of Plane Curves"

[49] [LCV] 6/25/05 Fine to Veblen

[50] [HUA] 9/7/13 Veblen to Birkhoff; [HUA] 12/25/13 Veblen to Birkhoff

[51] Archibald, *A Centennial History*, pp. 28–32

[52] Aspray, *Emergence*

[53] [IAS] 4/11/32 Trustees minutes

[54] [IAS] 4/11/32 Corporation minutes

[55] [IAS] 4/24/32 Veblen to Flexner

[56] Flexner, *Autobiography*, p. 251

[57] [IAS] 8/20/32 Flexner memorandum regarding Weyl

[58] [IAS] 4/24/32 Veblen to Flexner

[59] [IAS] 6/5/32 Veblen to Flexner; [IAS] 6/4/32 Flexner to Weyl

[60] [IAS] 8/20/32 Flexner memorandum regarding Weyl; [LCV] 6/28/32 Veblen to Miss Jones

[61] [IAS] 6/2/32 Flexner to Veblen; [IAS] 6/5/32 Veblen to Flexner; [IAS] 6/12/32 Veblen to Flexner

[62] Frank, *Einstein*

[63] Flexner, *Autobiography*, p. 252

[64] [IAS] 6/6/32 Flexner to Veblen

[65] [IAS] 6/6/32 Flexner to Einstein, Stern memo on terms

[66] [IAS] 6/8/32 Einstein to Flexner; [LCF] 6/14/32 Abe to Anne

[67] [IAS] 6/14/32 Flexner to Einstein

[68] [LCF] 6/14/32 Abe to Anne

[69] Stern, *History*, p. 137; [IAS] 6/14/32 Bailey to Bamberger and Fuld

[70] [LCV] 6/23/32 Eisenhart to Veblen

[71] [LCF] 6/29/32 Abe to Anne; [IAS] 6/30/32 Flexner to Veblen

[72] [IAS] 7/12/32 Flexner to Einstein

[73] [IAS] 7/30/32 Einstein to Flexner

[74] [IAS] 7/25/32 Millikan to Flexner; [IAS] 7/30/32 Flexner to Millikan; [APS] 8/4/32 Abe to Simon

[75] [IAS] 8/22/32 Hale to Flexner; [IAS] 8/30/32 Flexner to Hale; [APS] 11/28/32 Hale to Simon

[76] [IAS] 7/4/32 Weyl to Flexner

[77] [IAS] 7/16/32 Weyl to Flexner

[78] [IAS] 7/14/32 Aydelotte to Flexner

[79] [IAS] 7/16/32? Flexner cable to Weyl; [LCV] 7/17/32 Weyl to Veblen

[80] [IAS] 7/21/32 Flexner to Weyl; [IAS] 7/20/32 Born to Weyl; [IAS] 7/22/32 Flexner to Maass; [IAS] 7/28/32 Maass to Flexner

[81] [IAS] 7/30/32 Weyl to Flexner

[82] [IAS] 8/18/32 Flexner telegram to Bamberger; [IAS] 8/19/32 Bamberger telegram to Flexner

[83] [IAS] 8/23/32 Flexner to Weyl

[84] [IAS] 8/22/32 Flexner to Bamberger; [IAS] 8/20/32 Flexner memorandum regarding Weyl

[85] [IAS] 8/26/32 Bamberger to Flexner; [IAS] 9/21/32 Weyl to Flexner; [IAS] 10/7/32 Flexner to Weyl

[86] [IAS] 10/10/32 Trustees minutes

[87] [IAS] 8/30/32 Flexner to Einstein

CHAPTER SIX

[1] [IAS] 6/5/32 Veblen to Flexner

[2] [IAS] 6/30/32 Flexner to Veblen

[3] [IAS] 7/7/32 Veblen to Flexner

[4] Reference on Gödel: Dawson, *Logical Dilemmas*

[5] [IAS] 7/21/32 Flexner to Veblen

[6] [APS] 10/17/32 Simon to Abe; [APS] 10/24/32 Simon to Abe

[7] [APS] 10/21/32 Abe to Simon

[8] [APS] 10/27/32 Simon to Abe

[9] *New York Times* 10/11/32 pp. 1, 18; [IAS] 11/11/32 Veblen to Weyl

[10] [IAS] 11/25/32 Flexner to Veblen

[11] [IAS] 2/6/33 Flexner to Gödel; [LCV] 3/31/33 Gödel to Veblen

[12] [IAS] 10/12/32 Beard to Flexner; [IAS] 10/29/32 Frankfurter to Flexner

[13] [IAS] 11/7/32 Flexner to Frankfurter

[14] [IAS] 11/12/32 Flexner to Eisenhart

[15] [SMM] 12/3/32 Flexner to Eisenhart

[16] [IAS] 11/26/32 Eisenhart to Flexner

[17] [ABP] 12/17/32 Presented to the Curriculum Committee; [SMM] 1/8/32 Jacobus to Duffield

[18] [IAS] 11/11/32 Veblen to Weyl; [LCV] 11/7/32 Weyl to Veblen

[19] [LCV] 11/23/32 Veblen to Weyl; [IAS] 11/25/32 Flexner to Veblen

[20] [IAS] 11/25/32 Flexner to Weyl

[21] [IAS] 11/11/32 Veblen to Weyl

[22] [LCV] 12/1/32 Weyl to Flexner; [LCV] 12/5/32 Weyl to Veblen

[23] [LCV] 1/21/33 Courant to Veblen; Reference on Hitler and Nazi Party: Kershaw, *Hitler*

[24] [LCV] 12/1/32 Weyl to Flexner; [LCV] 12/5/32 Weyl to Flexner

[25] [LCF] 12/8/32 Abe to Jean

[26] [IAS] 7/4/32 Weyl to Flexner; [IAS] 8/23/32 Flexner to Weyl; [IAS] 12/7/32 Executive Committee minutes

[27] References on Alexander: Lefschetz, "James Waddell Alexander"; James, "Portrait"

[28] Gluchoff and Harmann, "On a Much Underestimated"

[29] Dieudonné, *History*

[30] [LCF] 12/22/32 Abe to Jean

[31] [LCV] 11/28/28 Fine to Veblen

[32] [LCV] 12/1/32 Weyl to Flexner; [IAS] 12/5/32 Flexner to Bamberger; [IAS] 12/7/32 Executive Committee minutes; [IAS] 12/7/32 Flexner to Weyl

[33] [IAS] 12/23/32 Maass to Flexner

[34] [IAS] 1/9/33 Trustees minutes

[35] [IAS] 12/7/32 Frankfurter to Flexner

[36] [IAS] 12/13/32 Flexner to Veblen

[37] References on Lefschetz: Archibald, *A Semicentennial History*, pp. 236–240; Lefschetz, "Reminisces"

[38] [SMM] see Leon Cohen interview and others in "The Princeton Mathematics Community in the 1930s: An Oral History Project," on the web at http://infoshare1.princeton.edu/libraries/firestone/rbsc/finding_aids/mathoral/math.html

[39] [IAS] 12/11/44 Veblen to Aydelotte; [IAS] 12/22/32 Flexner to Alexander; [IAS] 12/26/32 Alexander to Flexner

[40] Borel, "The School," p. 124; [IAS] 12/22/32 Flexner to Veblen

[41] [LCF] 12/22/32 Abe to Jean

[42] [IAS] 1/3,3,5/33 cables Weyl to Flexner

[43] [IAS] 1/3,?/33 cables Flexner to Weyl

[44] [IAS] 1/5/33 cable Bailey to Weyl

[45] [LCF] 1/21/33 Abe to Anne

[46] [LCV] 1/9/33 Flexner to von Neumann

[47] [LCV] 1/9/33 Flexner to Veblen; [LCV] 1/9/33 Flexner to Eisenhart; [LCF] 1/9/33 Abe to Anne

[48] [IAS] 1/9/33 Trustees minutes

[49] [IAS] 1/11/33 cable Weyl to Flexner; [LCV] 1/21/33 Courant to Veblen; [LCV] 4/30/33 von Neumann to Veblen

[50] [LCV] 1/13/33 Flexner to Weyl

[51] [ABP] 1/12/33 Eisenhart to the Curriculum Committee

[52] [LCV] 1/13/33 Flexner to Veblen

[53] [IAS] 1/16/32 Flexner to Aydelotte

[54] [LCF] 1/16/33 Abe to Anne

[55] [IAS] 1/28/33 Executive Committee minutes

[56] References on von Neumann: Aspray, *John von Neumann*; Halmos, "The Legend"; Macrae, *John von Neumann*

[57] Macrae, *John von Neumann*, p. 73

[58] [LCV] 8/21/29 Veblen to Eisenhart

[59] [LCV] 10/15/29 Veblen to von Neumann; [LCV] 11/13/29 von Neumann to Veblen; [LCV] 11/19/29 von Neumann to Veblen

[60] [SMM] 5/6/33 Special Committee of the Faculty to the Admin Committee; [SMM] 5/11/33 Administrative Committee minutes

[61] [IAS] 2/14/62 Leidesdorf to Oppenheimer, from 2/21/62 Faculty minutes

[62] [IAS] 3/1/32 Flexner to Bamberger

[63] Stern, *History*, p. 156

[64] Reference on economic situation: Kennedy, *Freedom from Fear*

[65] [IAS] 3/13/33 Veblen to Flexner; [IAS] 3/17/33 Veblen to Flexner

[66] [IAS] 3/27/33 Flexner to Veblen

[67] [LCV] 3/17/33 Flexner to Veblen; [IAS] 3/24/33 Veblen to Flexner; [IAS] 5/5/33 Flexner to Veblen

[68] [LCV] 1/21/33 Courant to Veblen

[69] [LCV] 2/2/33 Weyl to Veblen; [IAS] 2/17/33 Veblen to Flexner; [IAS] 2/24/33 Weyl to Flexner; [IAS] 3/11/33 Flexner to Weyl

[70] [LCV] 3/27/33 Weyl to Veblen

[71] [LCV] 4/12/33 Weyl to Veblen

[72] [IAS] 5/1/33 Flexner to Aydelotte

[73] [LCV] 4/24/33 Weyl to Veblen; [LCV] 4/25?/33 Weyl to Veblen

[74] Reference on Courant and Göttingen: Reid, *Courant*

[75] [LCV] 6/9/33 Weyl to Veblen, translation by Margrit Nash

[76] Kershaw, *Hitler*, p. 469

[77] [IAS] 7/10/33 Veblen to Flexner

[78] [IAS] 7/14/33 Flexner to Bamberger

[79] IAS] 7/19/33 telegram Bamberger to Flexner; [IAS] 7/24/33 Flexner to Bamberger; [IAS] 7/24/33 Flexner to Maass

[80] [LCF] 1/9/33 Abe to Anne; [IAS] 3/3/33 Flexner to Aydelotte; [IAS] 5/1/33 Flexner to Aydelotte

[81] [IAS] 8/1/33 Flexner to Bamberger; [IAS] 8/18/33 Aydelotte to Flexner; [IAS] 8/23/33 Flexner to Bamberger and Fuld

[82] [IAS] 9/7/33 Flexner to Weyl; [LCV] 9/8/33 Flexner to Veblen

[83] References on Einstein: Clark, *Einstein: The Life and Times*; Brian, *Einstein: A Life*; Frank, *Einstein: His Life*

[84] [IAS] 4/13/33 Flexner to Einstein

[85] [IAS] 3/26/33 Einstein to Flexner

[86] [IAS] 4/13/33 Einstein to Flexner

[87] [IAS] 4/26/33 Flexner to Einstein; [IAS] 7/29/33 Einstein to Flexner

[88] [IAS] 10/24/33 Flexner to Mayer

[89] [IAS] 8/8/33 Flexner to Einstein; [APS] 7/20/33 Abe to Simon; [LCV] 9/8/33 Flexner to Veblen

[90] [APS] 9/2/33 Abe to Simon; [APS] 9/25/33 Abe to Simon

[91] [APS] 10/18/33 Abe to Simon

[92] [IAS] 11/3/33 Flexner to President Franklin Roosevelt

[93] Clark, *Einstein*, pp. 513–514

[94] [IAS] 11/10/33 Einstein to Flexner; [IAS] 11/13/33 Flexner to Einstein; [IAS] 10/24/33 Flexner to Mayer

[95] [IAS] 11/14/33 Flexner to Maass

[96] [IAS] 11/15/33 Flexner to Einstein

[97] [IAS] 11/30/33 Hilb to Flexner; [IAS] 11/30/33 Einstein to Flexner; [IAS] 12/4/33 Flexner to Hilb; [IAS] 12/9/33 Einstein to Flexner

[98] Stern, *History*, p. 175; Porter, *From Intellectual Sanctuary*, pp. 419–422; [IAS] 12/11/33 Flexner to Bamberger; [APS] 12/11/33 Abe to Simon

[99] [LCV] 1/23/33 Flexner to Veblen; [LCV] 1/24/33 Veblen to Flexner; [IAS] 1/25/33 Flexner to Morse

[100] [IAS] 1/29/34 Trustees minutes

[101] Reference on Schrödinger: Moore, *A Life of Erwin Schrödinger*

[102] Moore, *A Life*, p. 128

[103] Stern, *History*, pp. 182–184

[104] [IAS] 6/25/34 Schrödinger to Dodds; [IAS] 6/25/34 Schrödinger to Flexner; [IAS] 7/4/34 Flexner to Schödinger, Stern notes; [IAS] 7/4/34 Flexner to Eisenhart, Stern notes

[105] [IAS] 5/5/34 Veblen to Morse; [LCV] 11/28/28 Fine to Veblen; [LCV] 12/28/28 Alexander to Veblen

[106] References on Morse: Pitcher, "Marston Morse"; Bott, "Marston Morse"

[107] [LCV] 11/7/32 Weyl to Veblen; [IAS] 7/4/32 Weyl to Flexner

[108] [IAS] 5/5/34 Veblen to Morse; [IAS] 6/10/34 Flexner to Morse; [LCV] 6/10/34 Flexner to Veblen; [IAS] 8/30/34 Flexner to Veblen; [IAS] 9/10/34 Veblen to Flexner; [LCV] 9/14/34 Flexner to Veblen; [IAS] 10/19/34 Morse to Graustein; [IAS] 10/24/34 Morse to Flexner; [IAS] 10/25/34 Flexner to Morse

CHAPTER SEVEN

[1] [IAS] 10/11/44 Director's report

[2] [LCF] 11/21/31 Abe to Jean

[3] [IAS] 11/7/32 Flexner to Frankfurter; [IAS] 12/7/32 Frankfurter to Flexner

[4] [IAS] 1/17/33 Flexner to Mitrany

[5] [IAS] 1/17/33 Flexner to Mitrany

[6] [IAS] 2/14/33 Flexner to Frankfurter

[7] [IAS] 2/23/33 Flexner to Frankfurter

[8] [IAS] 2/6/31 Mitrany to Flexner; [IAS] 1/14/33 Mitrany to Flexner

[9] [IAS] 1/23/33 Flexner to Mitrany

[10] [IAS] 1/14/33 Mitrany to Flexner

[11] [IAS] 1/17/33 Flexner to Mitrany; [IAS] 1/23/33 Flexner to Mitrany; [IAS] 1/25/33 Mitrany to Flexner

[12] [IAS] 2/13/33 Frankfurter to Flexner

[13] [IAS] 2/21/33 Flexner to Mitrany (2 letters); [IAS] 2/24/33 Mitrany to Flexner; [IAS] 4/24/33 Trustees minutes

[14] [APS] 3/12/32 Abe to Simon

[15] [IAS] 3/1/33 Flexner to Fuld; [IAS] 3/6/33 Flexner to Bamberger and Fuld; [IAS] 1/7/33 Frankfurter to Flexner

[16] [IAS] 11/14/33 Flexner to Maass; [IAS] 11/1/33 Flexner to Frankfurter

[17] [IAS] 12/11/33 Frankfurter to Flexner

[18] [IAS] 12/28/33 Flexner to Frankfurter

[19] [IAS] 1/4/34 Flexner to Frankfurter

[20] [IAS] 1/24/34 Frankfurter to Flexner

[21] [IAS] 1/24/34 Frankfurter to Flexner

[22] [IAS] 2/6/34 Flexner to Frankfurter

[23] [IAS] 2/21/34 Frankfurter to Flexner

[24] [IAS] 4/24/34 Frankfurter to Flexner

[25] [IAS] 3/21/34 Flexner to Aydelotte; [IAS] 3/21/34 Flexner to Frankfurter; [IAS] 4/2/34 Simon Flexner to Bernard Flexner; [IAS] 5/7/34 Flexner to Frankfurter

[26] [APS] 12/11/33 Flexner to Chickering; [IAS] 1/5/34 Mitrany to Flexner; [IAS] 1/22/34 Flexner to Mitrany

[27] [IAS] 1/29/34 Trustees minutes

[28] [IAS] 3/15/34 Flexner to Frankfurter; [IAS] 3/16/34 Flexner to Beatrice Earle; [IAS] 3/25/34 Earle to Flexner; [IAS] 4/2/34 Flexner to Stewart

[29] [IAS] 4/23/34 Trustees minutes

[30] [IAS] 3/31/34 Flexner to Earle; [IAS] 10/12/33 Flexner to Earle

[31] [IAS] 12/18/30 Flexner to Earle

[32] [IAS] 8/25/31? Earle to Flexner; [LCF] 2/3/32 Flexner to Bailey; [IAS] 4/23/34 Trustees minutes

[33] [IAS] 4/23/34 Trustees minutes

[34] [IAS] 4/23/34 Bamberger to the Trustees

[35] [SMM] 3/1/34 Dodds' memorandum of conversation with Morey; [SMM] 3/15/34 Morey to Dodds; [IAS] 3/29/34 Panofsky to Flexner

[36] [IAS] 2/21/34 Flexner to Frankfurter

[37] [IAS] 3/29/34 Panofsky to Flexner

[38] [IAS] 4/9/34 Morey to Flexner

[39] [IAS] 12/22/32 Flexner to Meritt; [IAS] 1/11/33 Flexner to Meritt

[40] [IAS] 6/28/34 Flexner to Morey

[41] [IAS] 5/7/34 Flexner to Lowe; [IAS] 6/10/34 Flexner to Morey; [IAS] 6/14/34 Morey to Flexner

[42] [IAS] 5/22/34 Flexner to Meritt; [LCF] 6/2/34 Abe to Jean; [IAS] 6/10/34 Flexner to Morey

[43] [IAS] 7/10/34 Flexner to Bamberger

[44] [IAS] 6/19/34 Flexner to Fuld and Bamberger; [IAS] 6/28/34 Flexner to Bamberger

[45] [IAS] 7/22/34 Flexner to Bamberger; [IAS] 8/30/34 Flexner to Veblen; [LCV] 9/14/34 Flexner to Veblen

[46] Reference on economic situation: Kennedy, *Freedom from Fear*

[47] [IAS] 9/2/34 Flexner to Lowe

[48] [IAS] 8/14/34 Lowe to Flexner; [IAS] 9/2/34 Flexner to Lowe

[49] [IAS] 1/22/34 Flexner to Mitrany; [IAS] 3/16/34 Mitrany to Flexner; [IAS] 4/25/34 Flexner to Mitrany; [IAS] 5/24/34 Flexner to Mitrany

[50] [IAS] 3/29/33 Flexner to Bamberger to Fuld

[51] [IAS] 10/8/34 Trustees minutes; [IAS] 11/6/34 Frankfurter to Flexner

[52] [IAS] 8/23/33 Aydelotte to Flexner

[53] [IAS] 10/8/34 Trustees minutes; [IAS] 10/9/34 Flexner to Earle

[54] [IAS] 6/10/34 Flexner to Veblen

[55] [IAS] 10/25/34 Flexner to Morse

[56] [IAS] 10/27/34 Flexner to Aydelotte

[57] [IAS] 4/29/34 Schumpeter to Flexner; [IAS] 11/2/34 Flexner to Frankfurter

[58] [IAS] 10/29/34 Frankfurter to Flexner

[59] [IAS] 10/30/34 Flexner to Frankfurter

[60] [IAS] 10/31/34 Frankfurter to Flexner; [IAS] 11/2/34 Flexner to Frankfurter; [IAS] 11/6/34 Frankfurter to Flexner

[61] [IAS] 9/19/34 Flexner to Earle

[62] [IAS] 11/8/34 Flexner to Frankfurter

[63] [IAS] 11/9/34 Frankfurter to Flexner

[64] [IAS] 11/12/34 Flexner to Frankfurter; [IAS] 11/13/34 Frankfurter to Flexner

[65] [IAS] 11/8/34 Flexner to Riefler; [IAS] 12/24/34 Flexner to Aydelotte; [IAS] 1/22/35 Frankfurter to Flexner

[66] [IAS] 11/16/34 Riefler to Flexner; [IAS] 11/16/34 Proposed Economic Unit memorandum by Riefler

[67] [IAS] 11/17/34 Flexner to Stewart

[68] [IAS] 11/19/34 Stewart to Flexner

[69] [IAS] 8/7/34 Earle to Flexner; [IAS] 11/24/34 Flexner to Earle; [IAS] 11/29/34 Earle to Flexner; [IAS] 11/30/34 Mitrany to Flexner; [IAS] 12/20/34 Mitrany to Flexner

[70] [IAS] 12/3/34 Earle to Flexner

[71] [IAS] 12/10/34 Earle to Flexner; [IAS] 12/3/34 Flexner to Earle; [IAS] 12/13/34 Flexner to Earle

[72] [IAS] 1/22/34 Flexner to Mitrany; [IAS] 3/16/34 Mitrany to Flexner; [IAS] 4/25/34 Flexner to Mitrany; [IAS] 5/14/34 Mitrany to Flexner; [IAS] 5/24/34 Flexner to Mitrany; [IAS] 11/30/34 Mitrany to Flexner; [IAS] 12/13/34 Flexner to Mitrany

[73] [IAS] 12/24/34 Flexner to Aydelotte

[74] [IAS] 1/14/35 Trustees minutes

[75] [IAS] 1/22/35 Frankfurter to Flexner

[76] [IAS] 1/15/35 Flexner to Riefler

[77] [IAS] 1/16/35 Frankfurter to Riefler

[78] [IAS] 1/19/35 Flexner to Frankfurter

[79] [IAS] 1/19/35 Flexner to Frankfurter

[80] [IAS] 1/19/35 Flexner to Riefler; [IAS] 1/21/35 Flexner to Riefler; [IAS] 2/5/35 Flexner to Aydelotte

[81] [IAS] 1/21/35 Frankfurter to Flexner; [IAS] 11/15/34 Aydelotte to Flexner; [IAS] 12/21/34 Aydelotte to Frankfurter; [IAS] 12/24/34 Flexner to Aydelotte

[82] [IAS] 2/18/35 Aydelotte to Frankfurter

[83] [IAS] 1/29/35 Weed to Frankfurter

[84] [IAS] 4/3/35 Frankfurter to Aydelotte

[85] [IAS] 3/26/35 Aydelotte to Leidesdorf; [IAS] 4/17/35 Aydelotte to Frankfurter

[86] [IAS] 4/22/35 Trustees minutes

[87] [IAS] 2/21/35 Flexner to Meritt; [IAS] 2/23/35 Meritt to Flexner; [IAS] 12/7/35 Flexner to Lowe; [APS] 11/21/30 Abe to Simon

[88] [IAS] 4/27/35 Flexner to Straus

[89] [IAS] 3/21/35 Panofsky to Flexner; [IAS] 3/31/35 Meritt to Flexner; [IAS] 5/2/35 Flexner to Lowe

[90] [IAS] 12/29/34 Panofsky to Flexner; [IAS] 1/9/35 Panofsky to Flexner; [IAS] 3/21/35 Panofsky to Flexner

[91] [IAS] 3/11/35 Flexner to Straus; [IAS] 3/18/35 Straus to Flexner; [IAS] 3/22/35 Flexner to Straus

[92] [IAS] 4/24/35 Flexner to Straus; [IAS] 4/26/35 Straus to Flexner; [IAS] 4/27/35 Flexner to Straus

[93] [IAS] 4/22/35 Trustees minutes

[94] [IAS] 4/24/35 Flexner to Bamberger; [IAS] 4/24/35 Flexner to Stewart

[95] [IAS] 5/1/35 Flexner to Mitrany; [IAS] 7/6/35 Flexner to Mitrany; [IAS] 5/2/35 Riefler to Flexner

[96] [IAS] 5/6/35 Flexner to Riefler

[97] [IAS] 5/8/35 Riefler to Flexner

CHAPTER EIGHT

[1] [IAS] 10/14/35 Trustees minutes

[2] [IAS] 10/28/35 Flexner to Bamberger

[3] [IAS] 10/29/35 Bamberger to Flexner

[4] [IAS] 10/30/35 Flexner to Bamberger; [IAS] 10/31/35 Flexner to Fuld

[5] [IAS] 4/22/35 Report of the Director

[6] [IAS] 1/31/35 Morey to Flexner

[7] [IAS] 6/6/35 Morey to Flexner

[8] [IAS] 9/30/35 Morey to Flexner

[9] [IAS] 11/11/35 Morey to Flexner

[10] [IAS] 4/9/34 Morey to Flexner

[11] [IAS] 1/22/36 Morey to Flexner

[12] [IAS] October 1935 Riefler to Flexner on Program of work; [IAS] 11/20/35 Flexner Memorandum for Riefler

[13] [IAS] 8/14/34 Lowe to Flexner; [IAS] 12/7/35 Flexner to Lowe; [IAS] 1/27/36 Trustees minutes; [IAS] 4/19/37 Trustees minutes

[14] [IAS] 12/3/35 Flexner to Earle

[15] [IAS] 1/27/36 Trustees minutes

[16] [IAS] 1/28/36 Flexner to Herzfeld

[17] Stern, *History*, p. 277

[18] [IAS] 1/28/36 Flexner to Herzfeld

[19] [IAS] 2/13/36 Flexner to Lowe

[20] [IAS] 1/31/36 Morey to Flexner; [IAS] 11/17/36 Morey to Panofsky; [IAS] 11/23/36 Morey to Flexner; [IAS] 1/25/37 Trustees minutes

[21] [IAS] 4/15/36 Bailey to Alexander; [IAS] 6/15/36 Executive Committee minutes

[22] [IAS] 10/23/35 Flexner to Mitrany; [IAS] 1/7/36 Flexner to Mitrany; [IAS] 1/13/36 Mitrany to Flexner; [IAS] 1/14/36 Flexner to Mitrany; [IAS] 1/18/36 Mitrany to Flexner

[23] [IAS] 5/23/36 Flexner to Earle

[24] [IAS] 3/13/36 Riefler to Flexner

[25] [IAS] 3/13/36 Riefler to Flexner

[26] [IAS] 6/15/36 Executive Committee minutes; [IAS] 11/13/39 Riefler to Aydelotte

[27] [IAS] 6/15/36 Executive Committee minutes

[28] [IAS] 10/13/36 Trustees minutes

[29] [IAS] 1/25/43 Report of Committee on Gest Library in Trustees minutes

[30] [IAS] 10/15/36 Trustees minutes

[31] [IAS] 2/25/36 Meritt to Flexner

[32] References on Goldman: [IAS] A Symposium in Memory of Hetty Goldman 1881–1972; http://www.mnsu.edu/emuseum/information/biography/fghij/goldman_hetty.html

[33] Flexner, *Autobiography*, p. 64; [IAS] 2/15/36 Flexner to Meritt; [IAS] 2/25/36 Meritt to Flexner; [IAS] 10/6/36 Meritt to Flexner

[34] [IAS] 10/13/36 Trustees minutes

[35] [IAS] 11/7/36 Flexner to Veblen; [IAS] 10/30/36 Flexner to Maass

[36] [IAS] 10/8/36 Goldman to Flexner; [IAS] 4/29/39 Goldman to Flexner

[37] [IAS] 5/21/37 J. Goldman to Flexner; [IAS] 5/22/37 Flexner to Leidesdorf; [IAS] 4/19/37 Trustees minutes

[38] Stern, *History*, pp. 286–287

[39] [IAS] 10/31/36 Flexner to Keppel

[40] [IAS] 11/17/36 Morey to Panofsky

[41] [IAS] 9/7/35 Flexner to Einstein; [LCV] 3/18/37 Flexner to Veblen

[42] [IAS] 3/19/37 Flexner to Fosdick

[43] [IAS] 4/13/37 Fosdick to Flexner

[44] [IAS] 7/3/37 Veblen to Flexner; [LCV] 7/9/37 Flexner to Veblen; [IAS] 6/9/37 Flexner to Morgenthau; [IAS] 6/17/37 Morgenthau to Flexner; [IAS] 7/28/37 Riefler to Flexner

[45] [IAS] 12/13/39 Riefler to Aydelotte

[46] [IAS] 4/19/37 Trustees minutes, p. 3; [IAS] 7/28/37 Riefler to Flexner; [IAS] 8/5/37 Flexner to Riefler; [IAS] 8/10/37 Flexner to Riefler; [IAS] 8/13/37 Riefler to Flexner

[47] [IAS] 8/5/37 Flexner to Leidesdorf

[48] [IAS] 8/6/37 Flexner draft of letter for Bamberger and Fuld wills

[49] [IAS] 8/5/37 Flexner to Riefler

[50] [IAS] 8/5/37 Flexner to Leidesdorf

[51] [IAS] 8/7/37 Flexner to Maass

[52] [IAS] 10/11/37 Trustees minutes

[53] [IAS] 10/11/37 Trustees minutes

[54] [IAS] 2/28/38 Flexner to Earle

[55] [FHL] 3/9/38 Flexner to Aydelotte

[56] [IAS] 1/24/38 Trustees minutes

[57] [IAS] 1/26/38 Flexner to Veblen

[58] [LCV] 2/14/38 Flexner to Veblen

[59] [IAS] 4/5/38 For Flexner (presumably from Stewart); [IAS] 3/30/38 Flexner to Veblen

[60] [IAS] 4/4/38 Flexner to Bamberger; [IAS] 4/6/38 Flexner to Bamberger

[61] [IAS] 4/18/38 Trustees meeting

[62] [IAS] 4/18/38 Trustees meeting

[63] [IAS] 6/27/38 Riefler to Flexner; [IAS] 7/14/38 Riefler to Flexner

[64] [IAS] in undated July 1938 letter from Stewart to Flexner

[65] [IAS] 7/16/38 Flexner to Bamberger

[66] [IAS] 7/19/38 Bamberger to Flexner

[67] [IAS] 7/28/38 Stewart to Flexner

[68] [IAS] 7/16/38 Flexner to Aydelotte

[69] [IAS] 7/14/38 Riefler to Flexner

[70] [IAS] 10/10/38 Trustees minutes

[71] [IAS] 10/4/38 Stewart to Flexner; [IAS] 10/13/38 Flexner to Stewart; [IAS] 8/27/38 Flexner to Maass; [IAS] 2/9/39 Flexner to Fosdick; [IAS] 11/25/39 Flexner to Aydelotte

[72] [LCV] 10/21/37 The Needs of the Mathematics Group in Fuld Hall; [IAS] 1/29/38 Flexner to Veblen; [IAS] 3/30/38 Flexner to Veblen; [IAS] 5/26/38 Aydelotte to Maass; [IAS] 5/28/38 Maass to Aydelotte

[73] [IAS] 6/1/38 Maass to Aydelotte; [IAS] 6/14/38 Maass to Flexner; [IAS] 6/24/38 Aydelotte to Flexner; [IAS] 7/1/38 Flexner to Veblen; [IAS] 7/25/38 Maass to Flexner; [IAS] 9/27/38 Flexner to Fuld and Bamberger

CHAPTER NINE

[1] [IAS] 6/28/39 Earle to Aydelotte

[2] [IAS] 1/28/36 Flexner to Herzfeld; [IAS] 2/2/38 Flexner to Goldman

[3] References on anti-Semitism at Princeton: Synnott, *The Half-Opened Door*; Porter, *From Intellectual Sanctuary*

[4] [LCV] 9/19/34 Alexander to Veblen

[5] [IAS] 6/9/39 Earle to Flexner

[6] Stern, *History*, p. 402; [IAS] 6/25/39 Earle to Aydelotte

[7] [IAS] 1/23/39 Trustees minutes

[8] [IAS] 1/23/39 Trustees minutes

[9] [IAS] 1/30/39 Flexner to Veblen; [IAS] 2/3/39 Mathematics minutes

[10] Stern, *History*, pp. 408–410

[11] [IAS] 2/10/39 Flexner to Veblen; [IAS] 2/13/39 Veblen to Flexner; [IAS] 2/14/39 Flexner to Veblen; [IAS] 11/25/39 Flexner to Aydelotte; [IAS] 2/4/39 Flexner to Fosdick

[12] [IAS] 2/9/39 Flexner to Fosdick; [IAS] 2/21/39 Fosdick to Flexner; [IAS] 4/13/44 Aydelotte to Hardin

[13] [IAS] 2/21/39 Aydelotte to Flexner; [IAS] 3/22/39 Flexner to Houghton

[14] [IAS] 3/15/39 Einstein, Goldman, and Morse to Flexner; [IAS] 3/15/39 draft of unsent letter from Warren to Einstein

[15] Clark, *Einstein*, p. 514

[16] [IAS] 3/15/39 Einstein, Goldman, and Morse to Flexner

[17] [IAS] 3/18/39 Flexner to Morse (unsent)

[18] [FHL] 3/26/39 Einstein to Aydelotte

[19] [IAS] 12/28/38 Flexner to Munro

[20] [FHL] 3/26/39 Einstein to Aydelotte; [FHL] 3/27/39 Mitrany to Aydelotte

[21] [IAS] 2/18/35 Aydelotte to Frankfurter

[22] [FHL] 4/11/39 Weed to Aydelotte

[23] [FHL] 4/11/39 Weed to Aydelotte

[24] [FHL] 3/26/39 Einstein to Aydelotte (mentions communication with Leidesdorf); [FHL] undated notes by Aydelotte

[25] [FHL] 5/4/39 Veblen to Aydelotte

[26] [IAS] 7/28/37 Riefler to Flexner; [IAS] 8/5/37 Flexner to Leidesdorf; [IAS] 8/10/37 Flexner to Riefler; [IAS] 8/13/37 Riefler to Flexner; [IAS] 8/18/37 Maass to Flexner; [IAS] 6/14/38 Maass to Flexner

[27] [IAS] Stern copies of 1/15/38 Aydelotte-Flexner correspondence

[28] [IAS] Stern copies of 1/15/38 Aydelotte-Flexner correspondence; [IAS] 1/27/39 Flexner to Aydelotte

[29] [IAS] 1/24/38 Trustees minutes

[30] [IAS] 5/9/39 Flexner to Riefler; [IAS] 5/14/39 Riefler to Flexner

[31] [IAS] 5/22/39 Trustees minutes

[32] [IAS] 5/22/39 Trustees minutes

[33] [IAS] 9/21/38 Flexner to Maass; [IAS] 6/14/38 Maass to Flexner

[34] [IAS] 6/18/39 Earle to Maass; [IAS] 7/26/39 Earle to Maass; [LCV] 8/9/39 Veblen to Einstein

[35] [IAS] 6/26/39 Flexner to Leidesdorf; [LCV] 8/9/39 Veblen to Einstein

[36] [IAS] 6/26/39 Flexner to Leidesdorf

[37] [IAS] 4/2/34 Simon to Ben Flexner

[38] [IAS] 6/9/39 Earle to Flexner

[39] [IAS] 7/21/39 Earle to Maass and Leidesdorf

[40] [IAS] 6/9/39 Earle to Flexner

[41] [IAS] 10/19/39 Flexner to Earle

[42] [IAS] 6/15/39 Maass to Earle

[43] [IAS] 6/20/39 Maass to Aydelotte; [IAS] 6/26/39 Veblen to Earle

[44] [IAS] 6/26/39 Veblen to Earle

[45] [LCF] 7/22/39 Abe to Ben; [IAS] 8/22/39 Flexner to Aydelotte

[46] [IAS] 12/23/32 Maass to Flexner; [IAS] 5/31/39 Maass to Flexner

[47] [IAS] 6/26/39 Flexner to Leidesdorf

[48] [FHL] 6/13/39 Earle to Aydelotte; [FHL] 6/15/39 Earle to Aydelotte; [LCV] 6/18/39 Earle to Maass; [FHL] 6/25/39 Earle to Aydelotte

[49] [IAS] 6/25/39 Earle to Aydelotte

[50] [IAS] 6/25/39 Earle to Aydelotte

[51] [FHL] 6/25/39 Earle to Aydelotte

[52] [IAS] 6/25/39 Earle to Maass; [IAS] 6/28/39 Earle to Aydelotte

[53] [LCV] 6/18/39 Earle to Maass; [IAS] 6/28/39 Earle to Aydelotte; [LCV] 6/28/39 Earle to Veblen

[54] [LCV] 8/9/39 Veblen to Einstein; [IAS] 7/17/39 Veblen to Aydelotte

[55] Ellis, *Founding Brothers*, p. 149

[56] [IAS] 7/13/39 Earle to Veblen; [FHL] 8/12/39 Anne Flexner to Marie Aydelotte

[57] [IAS] 7/13/39 Earle to Veblen; [IAS] 7/13/39 Earle to Aydelotte; [IAS] 7/17/39 Veblen to Aydelotte; [FHL] 7/19/39 Flexner to Aydelotte

[58] [LCV] 7/22/39 Earle to Veblen

[59] [IAS] 7/21/39 Earle to Maass with statement for Maass and Leidesdorf

[60] [IAS] 7/23/39 Earle to Maass; [IAS] 7/24/39 Maass to Earle; [IAS] 7/26/39 Earle to Maass

[61] [IAS] 8/2/39 Hardin to Flexner

[62] [FHL] 7/27/39 Maass to Aydelotte

[63] [IAS] 7/29/39 Flexner to Hardin

[64] [FHL] 8/12/39 Anne Flexner to Marie Aydelotte

[65] [IAS] 7/29/39 Flexner to Hardin

[66] [IAS] 8/2/39 Hardin to Flexner

[67] [IAS] 8/2/39 Hardin to Leidesdorf; [IAS] 8/2/39 Hardin to Maass; [LCV] 8/5/39 Earle to Veblen; [LCV] 8/9/39 Veblen to Einstein

[68] [IAS] 8/2/39 Einstein to Roosevelt

[69] [LCV] 8/13/39 Einstein to Veblen

[70] [LCF] 7/29/39 Abe to Jean; [LCF] 8/12/39 Abe to Jean

[71] [FHL] 8/12/39 Anne Flexner to Marie Aydelotte

[72] [IAS] 8/7/39 Flexner to Hardin

[73] [IAS] 8/9/39 Hardin to Flexner

[74] [LCV] 8/12/39 Flexner to Veblen

CHAPTER TEN

[1] [FHL] 8/29/39 Anne Flexner to Marie Aydelotte

[2] [FHL] 8/23/39 Flexner to Aydelotte; [IAS] 10/17/39 Flexner to Aydelotte

[3] [IAS] 8/18/39 Veblen to Earle; [LCV] 8/21/39 Earle to Veblen; [FHL] 8/24/39 Maass to Aydelotte; [FHL] 8/28/39 Flexner to Aydelotte; [IAS] 8/28/39 Einstein to Earle; [IAS] 9/6/39 Einstein to Earle

[4] [IAS] 8/31/39 Earle to Aydelotte; [FHL] 9/1/39 Earle to Aydelotte; [IAS] 10/1/39 Earle to Aydelotte

[5] [FHL] 8/30/39 Flexner to Aydelotte; [LCV] 9/4/39 Earle to Veblen

[6] [IAS] 10/2/39 Bamberger to Houghton; [IAS] 10/5/39 Houghton to Bamberger; [IAS] 10/9/39 Trustees minutes

[7] [IAS] 10/14/39 Houghton to Aydelotte

[8] [IAS] 10/12/39 Straus to Houghton

[9] [IAS] 11/15/39 Flexner to Aydelotte

[10] [IAS] 11/24/39 Faculty minutes

[11] [IAS] 11/24/39 Executive Committee minutes; [IAS] 11/25/39 Flexner to Aydelotte; [IAS] 5/13/40 Director's report

[12] [IAS] 10/11/44 Director's report

[13] [IAS] 1/16/40 Aydelotte to Leidesdorf; [IAS] 10/11/44 Director's report

[14] [IAS] 4/4/40 Thompson to Aydelotte; [IAS] 4/5/40 Aydelotte to Bamberger; [IAS] 4/13/44 Aydelotte to Hardin

[15] [IAS] 10/14/40 Director's report; [IAS] 12/22/42 Aydelotte to Leidesdorf; [IAS] 10/11/44 Director's report

[16] [IAS] 11/18/40 Aydelotte to Bamberger and Fuld; [IAS] 4/16/44 Director's report

[17] [IAS] 12/22/42 Aydelotte to Flexner

[18] [IAS] 11/30/42 Aydelotte to Bamberger

[19] [IAS] 12/8/42 Aydelotte to Bamberger; [IAS] 3/10/43 Aydelotte to Leidesdorf

[20] [IAS] 2/20/43 Bamberger will; [IAS] 10/11/44 Director's report

[21] [IAS] 2/25/43 Aydelotte to Leidesdorf; [IAS] 2/27/43 Aydelotte memorandum to chairman of Executive Committee; Stern, *History*

[22] [IAS] 7/21/43 Aydelotte to Moe; Stern, *History*

[23] [IAS] 12/14/43 Executive Committee minutes

[24] [IAS] 1/6/44 Aydelotte to Bamberger; [IAS] 10/11/44 Director's report .

[25] [IAS] 4/18/44 Trustees minutes

[26] [IAS] 10/11/44 Director's report; [FHL] 2/28/45 Aydelotte to Maass; Blanshard, *Frank Aydelotte*, p. 335

[27] [IAS] 1/17/44 Flexner to Aydelotte; [LCF] 6/19/44 Flexner to Leidesdorf

[28] [IAS] 10/11/44 Director's report; [IAS] 10/11/44 Flexner to Aydelotte; [IAS] 10/12/44 Panofsky to Aydelotte; [IAS] 10/17/44 Aydelotte to Flexner; Stern, *History*, p. 548

[29] [IAS] 11/20/44 Faculty minutes; Blanshard, *Frank Aydelotte*, p. 337

[30] [IAS] 1/19/45 Trustees minutes, report of Committee on Institute Policy; [IAS] 3/2/45 Trustees minutes (rough notes)

[31] [IAS] 1/29/45 Maass to Hardin; [FHL] 2/28/45 Aydelotte to Maass; [FHL] 3/2/45 Trustees minutes; [IAS] 3/2/45 Trustees minutes (rough notes)

[32] [IAS] 4/20/45 Trustees minutes and Director's report; [IAS] 4/27/45 Faculty minutes

[33] [IAS] 5/2/45 Earle to Aydelotte

[34] [IAS] 5/22/45 Faculty minutes

[35] [IAS] 6/5/45 Executive Committee minutes; [IAS] 10/19/45 Trustees minutes Director's report; [IAS] 12/18/45 Executive Committee minutes; [IAS] 10/18/46 Director's Report

[36] [LCV] 11/30/45 von Neumann to Veblen; Dawson, *Logical Dilemmas*

[37] [IAS] 12/13/45 Faculty minutes

[38] [IAS] 12/12/46 Alexander to Aydelotte

[39] [LCV] 5/21/43 von Neumann to Veblen

[40] [LCN] 3/24/45 Wiener to von Neumann; [LCN] 8/14/45 Harrison to von Neumann; [LCN] 8/19/45 von Neumann to Harrison; [LCN] 8/23/45 Harrison to von Neumann

[41] Louise Morse interview

[42] [IAS] 10/19/45 Trustees minutes; [LCN] 11/20/45 von Neumann to Harrison; [LCN] 11/20/45 von Neumann to Hutchins

[43] [IAS] 12/18/45 Trustees minutes; Blanshard, *Frank Aydelotte*, pp. 343–380; [IAS] 10/18/46 Director's report; [IAS] 11/19/46 Executive Committee minutes; [IAS] 4/21/47 Faculty minutes

[44] [IAS] 4/16/45 Moe to Flexner; [IAS] 4/20/45 Trustees minutes

[45] References on Strauss: Pfau, *No Sacrifice*; Strauss, *Men and Decisions*

[46] Stern, *History*

[47] Strauss, *Men and Decisions*, p. 271; Stern, *History*

[48] Reference on Oppenheimer: Bird and Sherwin, *American Prometheus*

[49] [IAS] 12/16/47 Trustees minutes

[50] [IAS] 11/21/39 Earle to Mitrany; [IAS] 5/10/49 Mitrany to Aydelotte; [IAS] 6/27/49 Mitrany to Aydelotte

[51] [IAS] 12/16/47 Trustees minutes

[52] [IAS] 12/16/47 Trustees minutes

[53] [IAS] 2/10/49 Trustees minutes

[54] Bird and Sherwin, *American Prometheus*

[55] Pfau, *No Sacrifice*; Strauss, *Men and Decisions*

[56] Pfau, *No Sacrifice*, pp. 106–109

[57] Stern, *History*

[58] Bird and Sherwin, *American Prometheus*

[59] [IAS] 5/6/49 Oppenheimer to Mitrany; [IAS] 5/10/49 Mitrany Note for Oppenheimer; [IAS] 5/12/49 Mitrany to Oppenheimer

[60] Bird and Sherwin, *American Prometheus*

[61] [LCN] 3/15/55 von Neumann to Oppenheimer; [LCN] 12/22/55 Montgomery to von Neumann

[62] Macrae, *John von Neumann*, p. 373; [LCF] 3/19/56 von Neumann to Sproul

[63] 2/27/2003 email Borel to Batterson

[64] Borel, "The School of Mathematics"; [IAS] 4/6/62 Trustees minutes

[65] [IAS] 2/21/62 Faculty minutes; [IAS] 4/6/62 Trustees minutes

[66] Borel, "The School of Mathematics," p. 139

[67] Armand Borel interview

[68] Bird and Sherwin, *American Prometheus*, p. 581; Borel, "The School of Mathematics," pp. 138–139; [IAS] 10/27/55 Trustees minutes

[69] [IAS] Report of the Director 1966–1976; [IAS] Erica Mosner summary of materials from Committee on the Future; Borel interview; White, *Philosopher's Story*, p. 293

[70] Halberstam, *The Best and the Brightest*

[71] 3/2/73, 3/4/73, 3/25/73, 4/28/73, 4/29/73, 2/28/74 *New York Times* articles by Israel Shenker

[72] 4/28/73 *New York Times*, p. 29

[73] Landon Jones, *Bad Days on Mount Olympus*, p. 51

[74] Atle Selberg interview

[75] Phillip Griffiths interview

[76] IAS Report for the Academic Year 2003–2004; Peter Goddard interview

[77] Raoul Bott, "Marston Morse"

[78] Jackson, "The IAS School"

[79] Feynman, *"Surely You're Joking, Mr. Feynman!"* p. 165

[80] Regis, *Who Got Einstein's Office?* p. 280

[81] [IAS] 10/10/30 edited Director's report

[82] [IAS] 3/27/33 Flexner to Veblen; [LCF] 9/27/45 Hildebrandt to Flexner; [LCF] 10/5/45 Flexner to Hildebrandt; Borel interview; IAS Procedures for Academic Governance of the Institute (revised May 2002)

BOOKS, ARTICLES, AND OTHER NON-ARCHIVAL SOURCES

Albers, Don, Alexanderson, G. L., and Reid, Constance. *International Mathematical Congresses*. Springer-Verlag, New York, NY (1986).

Archibald, Raymond. *A Semicentennial History of the American Mathematical Society 1888–1938*, Volume I. American Mathematical Society, Providence, RI (1938).

Aspray, William. "The Emergence of Princeton as a World Center for Mathematical Research, 1896–1939" in *A Century of American Mathematics*, Part II, edited by P. Duren. American Mathematical Society, Providence, RI (1989), 195–215.

Aspray, William. *John von Neumann and the Origins of Modern Computing.* MIT Press, Cambridge, MA (1990).

Barrow, Clyde. *More than an Historian: The Political and Economic Thought of Charles A. Beard.* Transaction Publishers, New Brunswick, NJ (2000).

Bennett, Helen. "Do the Wise Thing If You Know What It Is—But Anyway Do Something!" *The American Magazine* 95 (1923), 72–122.

Bird, Kai and Sherwin, Martin. *American Prometheus: The Triumph and Tragedy of J. Robert Oppenheimer.* Knopf, New York. NY (2005).

Birkhoff, Garrett. "Mathematics at Harvard, 1836–1944" in *A Century of American Mathematics*, Part II, edited by P. Duren. American Mathematical Society, Providence, RI (1989), 3–58.

Birkhoff, Garrett. "Some Leaders in American Mathematics: 1891–1941" in *The Bicentennial Tribute to American Mathematics*, edited by D. Tarwater. Mathematical Association of America, Washington, DC (1977), 25–78.

Blanshard, Frances. *Frank Aydelotte of Swarthmore.* Wesleyan University Press, Middletown, CT (1970).

Bonner, Thomas. *Iconoclast: Abraham Flexner and a Life in Learning.* Johns Hopkins University Press, Baltimore, MD (2002).

Borel, Armand. "The School of Mathematics at the Institute for Advanced Study" in *A Century of American Mathematics*, Part III, edited by P. Duren. American Mathematical Society, Providence, RI (1989), 119–147.

Bott, Raoul. "Marston Morse and His Mathematical Works." *Bulletin American Mathematical Society* New Series 3 (1980), 907–950.

Breit, Edward and Hirsch, Barry (Editors). *Lives of the Laureates.* MIT Press, Cambridge, MA (1986).

Brian, Denis. *Einstein: A Life*. John Wiley, New York, NY (1996).

Chernow, Ron. *The Warburgs*. Random House, New York, NY (1993).

Chernow, Ron. *Titan: The Life of John D. Rockefeller, Sr.* Vintage, New York, NY (1999).

Clark, Ronald W. *Einstein: The Life and Times*. World Publishing, River Edge, NJ (1971).

Corner, George. *A History of the Rockefeller Institute*. Rockefeller Institute Press, New York, NY (1964).

Dawson, John W. *Logical Dilemmas: The Life and Work of Kurt Gödel*. A K Peters, Natick, MA (1997).

Diacu, Florin and Holmes, Philip. *Celestial Encounters: The Origins of Chaos and Stability*. Princeton University Press, Princeton, NJ (1996).

Dieudonné, Jean. *A History of Algebraic and Differential Topology 1900–1960*. Birkhäuser, Boston, MA (1989).

DuBois, W. E. B. *The Autobiography of W. E. B. DuBois*. International Publishers, New York, NY (1968).

Ellis, Joseph. *Founding Brothers: The Revolutionary Generation*. Knopf, New York, NY (2000).

Feynman, Richard. *"Surely You're Joking, Mr. Feynman!": Adventures of a Curious Character*. Norton, New York, NY (1985).

Flexner, Abraham. "A Modern University." *Atlantic Monthly* 136 (1925), 530–541.

Flexner, Abraham. *Abraham Flexner: An Autobiography*. Simon and Schuster, New York, NY (1960).

Flexner, Abraham. *Daniel Coit Gilman*. Harcourt, Brace, and Company, New York, NY (1946).

Flexner, Abraham. *Medical Education in the United States and Canada*, Bulletin Number Four. The Carnegie Foundation for the Advancement of Teaching, Stanford, CA (1910).

Flexner, Abraham. *Universities: American, English, German*. Oxford University Press, Oxford, UK (1930).

Flexner, James. *An American Saga: The Story of Helen Thomas and Simon Flexner*. Little, Brown, and Company, Boston, MA (1984).

Fosdick, Raymond. *The Story of the Rockefeller Foundation*. Harper, New York, NY (1952).

Frank, Philipp. *Einstein: His Life and Times*. Knopf, New York, NY (1967).

French, John. *A History of the University Founded by Johns Hopkins*. The Johns Hopkins Press, Baltimore, MD (1946).

Franklin, Fabian. *The Life of Daniel Coit Gilman*. Dodd, Mead, and Company, New York, NY (1910).

Gluchoff, Alan and Harmann, Frederick. "On a 'Much Underestimated Paper' of Alexander." *Arch. Hist. Exact Sciences* 55 (2000), 1–41.

Goodspeed, Thomas. *A History of the University of Chicago*. University of Chicago Press, Chicago, IL (1916).

Halberstam, David. *The Best and the Brightest*. Random House, New York, NY (1972).

Halmos, P. "The Legend of John Von Neumann." *American Mathematical Monthly* 80 (1973), 382–394.

Hawkins, Hugh. *Pioneer: A History of the Johns Hopkins University, 1874–1889*. Cornell University Press, Ithaca, NY (1960).

Hawkins, Thomas. *Emergence of the Theory of Lie Groups: An Essay in the History of Mathematics 1869–1926*. Springer-Verlag, New York, NY (2000).

Institute for Advanced Study. *A Community of Scholars: The Institute for Advanced Study Faculty and Members 1930–1980*. The Institute for Advanced Study, Princeton, NJ (1980).

Institute for Advanced Study. *Annual Report for the Academic Year 2003–2004.* The Institute for Advanced Study, Princeton, NJ (2004).

Jackson, Allyn. "The IAS School of Mathematics." *Notices of the American Mathematical Society* 49 (September 2002), 896–904.

James, Ioan M. "Portrait of Alexander." *Bulletin American Mathematical Society* New Series 38 (2001), 123–129.

Jones, Landon Y. "Bad Days on Mount Olympus." *The Atlantic* 37 (February 1974), 37–53.

Kennedy, Donald. *Freedom from Fear: The American People in Depression and War 1929–1945.* Oxford University Press, Oxford, UK (1999).

Kershaw, Ian. *Hitler 1889-1936: Hubris.* Norton, New York, NY (1998).

Lefschetz, Solomon. "James Waddell Alexander (1888–1971)" in *American Philosophical Society Yearbook 1973.* American Philosophical Society, Philadelphia, PA (1974).

Lefschetz, Solomon. "Reminiscences of a Mathematical Immigrant in the United States" in *A Century of American Mathematics,* Part I, edited by P. Duren. American Mathematical Society, Providence, RI (1989), 201–207.

Link, William. *The Paradox of Southern Progressivism, 1880–1930.* University of North Carolina Press, Chapel Hill, NC (1992).

Macrae, Norman. *John von Neumann.* Pantheon, New York, NY (1992).

Meltzer, Allan. *A History of the Federal Reserve Volume I, 1913–1951.* University of Chicago Press, Chicago, IL (2003).

Montgomery, Deane. "Oswald Veblen" in *A Century of American Mathematics,* Part I, edited by P. Duren. American Mathematical Society, Providence, RI (1989), 118–129.

Moore, Walter. *A Life of Erwin Schrödinger.* Cambridge University Press, Cambridge, UK (1994).

New York Times, 30 June 1929, 22 August 1929, 12 December 1930, 11 October 1932, 2 March 1973, 4 March 1973, 25 March 1973, 28 April 1973, 29 April 1973, and 28 February 1974.

Newman, M. H. A. "Hermann Weyl." *Biographical Memoirs of Fellows of the Royal Society* 3 (1957), 305–328.

Pais, Abraham. *Einstein Lived Here*. Oxford University Press, Oxford, UK (1994).

Pais, Abraham. *'Subtle is the Lord...': The Science and the Life of Albert Einstein*. Oxford University Press, Oxford, UK (1982).

Parshall, Karen and Rowe, David. *The Emergence of the American Mathematical Research Community 1876–1900: J. J. Sylvester, Felix Klein, and E. H. Moore*. American Mathematical Society, Providence, RI (1994).

Pfau, Richard. *No Sacrifice Too Great: The Life of Lewis L. Strauss*. University Press of Virginia, Charlottesville, VA (1984).

Pitcher, Everett. "Marston Morse." *National Academy of Sciences Biographical Memoirs* 65 (1994), 223–240.

Porter, Laura. "From Intellectual Sanctuary to Social Responsibility: The Founding of the Institute for Advanced Study, 1930–1933." PhD thesis, Princeton University (1988).

Regis, Ed. *Who Got Einstein's Office?* Addison-Wesley, Reading, MA (1987).

Reid, Constance. *Courant in Göttingen and New York*. Springer-Verlag, New York, NY (1976).

Reid, Constance. *Hilbert*. Springer-Verlag, New York, NY (1970).

Reid, Constance. "The Road Not Taken: A Footnote in the History of Mathematics." *The Mathematical Intelligencer* 1(1) (1978), 21–23.

Richardson, R. G. D. "The PhD Degree and Mathematical Research." *American Mathematical Monthly* 43 (1936), 199–215.

Shils, Edward (Editor). *Remembering the University of Chicago: Teachers, Scientists, and Scholars*. University of Chicago Press, Chicago, IL (1991).

Siegmund-Schultze, Reinhard. *Rockefeller and the Internationalization of Mathematics Between the Two World Wars*. Birkhäuser, Boston, MA (2001).

Sigurdsson, Skuli. "Hermann Weyl, Mathematics and Physics, 1900–1927." PhD thesis, Harvard University (1991).

Stern, Beatrice. *A History of the Institute for Advanced Study 1930–1950*. Unpublished manuscript.

Storr, Richard. *Harper's University: The Beginnings*. University of Chicago Press, Chicago, IL (1966).

Strauss, Lewis L. *Men and Decisions*. Doubleday, Garden City, NY (1962).

Synnott, Marcia. *The Half-Opened Door*. Greenwood Press, Westport, CT (1979).

Veblen, Oswald. "George David Birkhoff." *Yearbook of the American Philosophical Society* (1946), 279–285.

Veblen, Oswald. "Theory of Plane Curves in Non-metrical Analysis Situs." *Transactions American Mathematical Society* 6 (1905), 83–98.

Weyl, Hermann. "David Hilbert and His Mathematical Work." *Bulletin American Mathematical Society* 50 (1944), 612–654.

White, Morton. *A Philosopher's Story*. Pennsylvania State University Press, University Park, PA (1999).

INTERVIEWS

Borel, Armand 2/11/00
Goddard, Peter 3/16/05
Griffiths, Phillip 3/15/05
Morse, Louise 7/13/99
Selberg, Atle 3/15/05
White, Morton 3/16/05

INDEX

Albert, Adrian, 153
Albright, William, 248
Alexander polynomial, 129
Alexander's Chimney, 129
Alexander, Elizabeth Alexander, 128
Alexander, James W., I, 128
Alexander, James W., II, 59, 106, 107,
    119, 120, 123, 127–133,
    155, 157, 189, 197, 216,
    218, 232, 249, 263, 264
Alexander, John White, 128
Alexandroff, Pavel, 119
algebraic topology, 106, 128, 156
allopathy, 23
American Journal of Mathematics, 9
American Mathematical Society, 21,
    98, 106
analysis situs, 106, 107, 128
Antioch, 170, 192–194, 197, 204,
    216
Artin, Emil, 109, 110, 115, 119, 120
Atomic Energy Commission (AEC),
    252, 254–255
Aydelotte, Frank, 50, 55, 56, 70, 76,
    79, 115, 130, 137, 146,
    148, 165, 178, 183, 191,
    207, 213, 219, 259
    director, 239–251
    role in Flexner coup, 223–237

bacteriology, 8, 9, 23
Bahcall, John, 263
Bamberger's (store), 33, 34
Bamberger, Edgar, 49, 76, 96, 143,
    245, 247

Bamberger, Lavinia, 227
Bamberger, Louis, ix, 31–53, 56, 67,
    70, 75, 76, 78, 86, 88,
    95–97, 100, 108, 112,
    114–116, 130–131, 143,
    144, 147–148, 152, 159,
    162, 165, 168, 173, 175,
    181, 185, 186, 195, 196,
    198, 204–213, 215, 218,
    222, 225, 227, 230–237,
    239–245, 250, 251, 264
    suspends support, 190–192
Beard, Charles, 64, 73, 79, 94, 122,
    123, 159, 164, 167
Bellah, Robert, 259
Betti numbers, 128, 156
Beurling, Arne, 256, 264
Birkhoff, Garrett, 101
Birkhoff, George D., 21, 56, 59–66,
    70, 73, 75, 76, 78, 79, 81,
    88, 89, 94, 95, 102, 105,
    107, 115, 140, 143, 148,
    155, 174, 187, 192, 196
    biography, 61–63
    negotiation, 75, 96–101
Bôcher, Maxime, 11, 20, 61, 62
Bohr, Niels, 205–206, 208, 253
Bolza, Oskar, 11, 20, 21
Bombieri, Enrico, 264
Borel, Armand, 256–257, 259, 264
Bott, Raoul, 262
Bourgain, Jean, 264
Brandeis, Louis, 57, 58
Brauer, Richard, 157

Bronk, Detlev, 252
Bryn Mawr College, 200
Buttrick, Wallace, 26, 29, 52

California Institute of Technology, 81,
        86–88, 111, 114, 148
Campbell, W. A., 194, 196, 197
Capps, Edward, 201–202
Caputh, 110, 114
Carnegie
        Corporation, 204, 234
        Foundation, 26
Carrel, Alexis, 25, 49
Cherniss, Harold, 253
Church, Alonzo, 157
Churchill, Winston, 25
Clark University, 19, 20, 132
Clark, Maurice, 16
Clay, Henry, 210, 212
Columbia University, 37, 43–45, 56,
        57
Conant, James, 157
Coolidge, Julian, 98–100
Cornell University, 37
        Medical School, 28
Courant, Richard, 146–147
Crawford, Anne, *see* Flexner, Anne
        Crawford

de Tolnay, Charles, 246
Deligne, Pierre, 264
Dickson, Leonard, 21, 56, 59
Dilworth, J. Richardson, 259, 260
Dinsmoor, William, 171, 193, 194
Dirac, Paul, 119, 140, 144, 155, 157,
        175, 182, 197, 205–206,
        253
Dirichlet, Johann P. G. L., 11
Dodds, Harold, 141, 169, 184, 216,
        227, 229
Douglas, Jesse, 157
Duffield, Edward, 143
Dyson, Freeman, 256

Earle, Edward, 174, 175, 180, 187,
        190, 198, 205, 207, 211,
        240, 253, 255
    hiring, 166–167, 174
    letter with faculty grievances, 229
    role in Flexner coup, 215–216,
        221–223, 228–236
Eastman, George, 28, 37
Eidgenössische Technische
        Hochschule (ETH), 82, 83,
        92, 94, 139, 248
Einstein, Albert, x, 1, 92, 94, 95, 98,
        102, 107, 114–116, 122,
        134, 145, 154, 155, 157,
        187, 189, 197, 204,
        210–212, 227, 240,
        243–244, 246, 256, 264
    biography, 81–87
    conflict with Flexner, 117,
        148–152, 219–221,
        223–224, 228–237
    hiring, 81, 86–90, 108, 110–113
Eisenhart, L. P., 68, 69, 77, 106, 110,
        113, 123–125, 127,
        131–133, 135–137,
        139–141, 154–155, 216
Eisenhower, Dwight, 255
Eliot, Charles, 12
epigraphy, 171

Federal Reserve, 162, 179, 198
        Bank, 57
        Board, 211, 254
Feynman, Richard, 251, 263
Fields Medal, x, 262, 264
Fine Hall, 68, 107, 113, 123,
        153–155, 174, 189, 190,
        203, 213, 218, 225
Fine Professor, 68, 131, 133
Fine, Henry B., 11, 61–62, 68, 104,
        106, 113, 133

Flexner, Abraham, ix–xi, 1, 31, 66, 86,
        142, 215, 239–248, 251,
        254, 255, 258, 261–262,
        264
    attempt to control Einstein,
        150–152
    biography, 2, 6, 8, 12–13
    budget, 71, 95, 109, 119, 129,
        168, 174, 175, 184, 185,
        189–213, 216, 218–219,
        227, 228, 242, 244
    choosing first subject, 51, 55–66,
        72–73, 78, 102
    considering economics, 57,
        63–66, 73–75, 79, 94–95,
        122–123, 131, 159–163,
        167, 172–174, 176, 184,
        187, 194, 198–199,
        205–212
    considering humanities, 58,
        167–172, 174, 184–186,
        192–194, 200–204
    considering mathematics, 56,
        59–66, 78, 88–90, 94–95,
        127–129, 131–137, 143,
        146–148, 154–157,
        204–205
    discrimination, 38, 74, 90, 127,
        131, 144, 177, 216–219,
        225–226, 229, 235, 264
    establishing Institute for
        Advanced Study, 43–53
    first offers, 95–116
    full-time approach, 27, 28, 71,
        88, 97, 111, 196
    General Education Board,
        26–29, 68, 102, 164, 199,
        230
    Gilman's influence, 6, 9, 12, 39,
        43, 55, 97, 169, 187

    lack of scholarship, 13, 43, 57,
        102, 178, 241
    medical education, ix, 26–29, 36,
        37
    on genius, 52–53, 56, 60, 81,
        187, 209, 264
    pitching graduate university,
        36–42
    retirement, 53, 166, 215–237
    Veblen's influence, 69, 77–78,
        102–103, 119–122,
        125–129, 134–137, 147,
        152–154
Flexner, Anne Crawford, 12, 111,
        228, 234, 236–237, 239
Flexner, Eleanor, 64, 66
Flexner, Jacob, 2, 7, 12
Flexner, Jean, 237
Flexner, Simon, 7–9, 12, 13, 24–26,
        29, 37, 43, 49, 51, 53, 68,
        75, 101, 120–121, 130,
        165, 228, 231
Frank, Louis, 32, 34
Frankfurter, Felix, 60, 64, 73, 75, 76,
        79, 122–123, 130, 131,
        160–163, 169, 175, 176,
        179, 186, 215, 223
    disputes with Flexner, 163–166,
        176–178, 181–183
Freedman, Michael, 262
Friedenwald, Julius, 49
Friedman, Milton, 64
Frisch, Ragnar, 66
Fuld Hall, 207, 212–213, 218, 222,
        225, 227, 242, 249
Fuld, Carrie, ix, 31, 32, 34–36, 39,
        41–43, 49, 51, 67, 86,
        95–97, 100, 112, 130, 144,
        162, 186, 190, 192, 206,
        207, 235, 239, 243, 245,
        250, 251, 264
Fuld, Felix, 32, 34–36, 39

Gates, Frederick T., 17–19, 22–29, 37, 48
Gauss, Carl Friedrich, 1, 11, 93
Geertz, Clifford, 259
General Advisory Committee, 252, 254
General Education Board, *see also* Flexner, Abraham, General Education Board, 26–29, 37
German Education Ministry, 114
Gest Library, 199–200, 208
Gest, G. M., 199
Gildersleeve, Basil, 6, 53, 55
Gilman, Daniel Coit, 2–13, 15, 18, 19, 21, 26, 38, 39, 41, 43, 48, 52, 55, 65, 79, 89, 95, 121, 178, 186, 191, 229
Goddard, Peter, 261
Gödel, Kurt, 120–122, 140, 153, 187, 204, 208, 249, 251, 256, 264
Goldberger, Marvin, 260
Goldman, Hetty, 200–203, 213, 215, 219–221, 227, 228, 232, 244, 253
Goldman, Julius, 200, 202
Göttingen, 1, 10, 11, 21, 63, 66, 89, 90, 93, 103, 105, 109, 115, 126, 139, 249
Great Depression, xi, 1, 42, 95, 173, 190
Griffiths, Phillip, 260–261
Gronwall, T. H., 128
Grothendieck, Alexander, 256

Hadamard, Jacques, 65, 66
Hale, William, 18
Hardin, Charles, 52
Hardin, John, 33, 34, 45, 49, 52, 78, 130, 222, 239, 240, 245, 247, 251
    role in Flexner coup, 230–237

Harish-Chandra, 256, 258
Harper, William Rainey, 17–22, 24, 44, 48, 196
Harvard University, 4, 12, 13, 18, 20, 21, 28, 43, 44, 59, 60, 62, 63, 78, 97, 98, 161, 162, 197
Hebrew University, 86
Heisenberg, Werner, 139
Herzfeld, Ernst, 193–197, 200–203, 208, 213, 215, 232, 243–244, 248
Hibben, John Grier, 69, 77, 124, 141, 143
Hilbert, David, 1, 11, 90–94, 105, 126, 139
Hirschman, Albert, 260
Hitler, Adolf, xi, 1, 85–87, 94, 102, 103, 108, 126, 137, 144, 147, 186, 237, 254
homeopathic medicine, 23, 31
Hoover, Herbert, 43
Hopkins, Johns, 2, 15, 35, 186
Hormander, Lars, 258
Houghton, Alanson, 50, 130, 219, 230, 239, 241, 245
Husserl, Edmund, 92
Hutchins, Robert, 184

Institute for Advanced Study
    biology, 261
    commitment to establish, 42
    Committee on Institute Policy, 245–247
    computer science, 261
    East Asian studies, 261
    endowment, 44, 48, 131, 168, 196, 227, 242, 243, 245, 260, 261
    Latin-American studies, 240, 242, 243
    letter to charter trustees, 45–48

location, 39, 41, 45, 48, 52, 66–67, 69–70, 75–77, 96–97, 107, 114, 123

naming, 44–45, 47

nondiscriminatory ideals, 36, 40, 41, 46, 74, 90, 127, 144, 177, 201, 264

Oriental (Chinese) studies, 242, 243, 250, 251, 253

postdoctoral program, evolution of, x, 3, 4, 6, 12, 20, 21, 38–40, 46, 47, 121–122, 152–154, 261

Princeton faculty policy, 78, 107, 109, 113, 120, 123–125, 127–128, 131–133, 135–137, 141–144, 256–258

School of Economics and Politics, 162, 167, 174, 183, 189, 211–212, 248, 253, 254

School of Historical Studies, 254, 255

School of Humanistic Studies, 169, 171, 174, 184, 186, 189, 193–195, 200, 202, 248, 251, 253, 254

School of Mathematics, x, 1, 66, 119, 157, 174, 189, 208, 248, 253, 254, 256–258, 261, 264

School of Natural Sciences, 258

School of Social Science, 258–260

theoretical physics, 144, 154–258

trustee selection, 49, 51, 74, 78, 162, 163, 175, 183, 251

International Education Board, 29, 59

Johns Hopkins University, 2, 4–13, 18, 19, 23, 25, 27, 37, 38, 43, 48, 53, 60, 121, 122, 134, 184, 241

Jones, Landon, 259

Jones, Thomas, 68

Jordan Curve Theorem, 104

Joseph, Hella, see Weyl, Hella Joseph

Kaysen, Carl, 258–261

Kennan, George, 255–256

Klein, Felix, 9–12, 21, 61, 84, 90, 105, 126

Laughlin, J. Laurence, 18

Lee, T. D., 256, 258

Lefschetz Fixed Point Theorem, 132

Lefschetz, Solomon, 59, 65, 76, 107, 119, 120, 123, 131, 133, 155, 156, 174, 216, 263

Lehman, Herbert, 49, 175

Leidesdorf, Samuel, 36–39, 42, 49, 95, 96, 116, 130, 137, 142–143, 175, 184, 206, 239–240, 245, 247, 251, 257

role in Flexner coup, 222–237

Lemaître, Georges, 157

Leontief, Wassily, 66

Les Houches, 129

Levitzkaya, Natalie, 129

Library of Congress, 57, 199

Lowe, Elias A., 169–172, 174, 175, 180, 184, 185, 190, 193–196, 200–203, 232, 243–244, 249

Lowell, A. Lawrence, 28, 98–100, 102, 157

Löwenthall, Elsa, 84

Lowndes, Beatrice (Bee), 167

Maass, Herbert, 36–39, 42, 49, 70, 76, 78, 88, 95, 96, 116,

130–131, 137, 143, 151,
152, 175, 191, 202, 206,
207, 213, 239–240, 255
role in Flexner coup, 222–237
wields power, 245–252
Macy's, 35, 36, 49, 241
Manhattan Project, 250, 252, 254
Maric, Mileva, 82, 84
Marschak, Jacob, 176–178, 180
Maschke, Heinrich, 11, 20, 21
Massachusetts Institute of Technology,
250
Mayer, Walther, 111, 112, 116–117,
120, 149, 152, 189, 211
Mayer-Vietoris sequence, 117
McCormick, John Rockefeller, 24
Meiss, Millard, 257
Meritt, Benjamin, 171, 174, 184–186,
189, 190, 193, 194,
200–202, 221, 228, 232,
253, 255, 258
Michelson, Albert A., 19
Millikan, Robert, 86, 114
Milnor, John, 142, 256–258
Minkowski, Hermann, 90
Mitrany, David, 161–162, 167, 174,
180, 181, 187, 189, 190,
198, 203, 205, 207, 211,
215, 218, 221–223,
230–232, 253, 255
Mitrany, Ena, 161
Mittag-Leffler, Gösta, 105
Moe, Henry, 251, 252
Montgomery, Deane, 157, 254, 256,
264
Moore, Eliakim Hastings (E. H.),
19–21, 59, 61, 104, 105
Moore, R. L., 104
Morey, Charles, 58, 168–169, 177,
185–187, 190, 192–194,
196, 197, 201, 203, 204,
215
memorandum, 169–172

Morgan, J. P., 28
Morgan, Thomas, 88
Morison, Samuel Eliot, 248, 251
Morse Theory, 156, 262
Morse, Marston, 59, 119, 154–157,
172, 174–176, 189, 190,
196, 197, 219–221, 227,
232, 249, 251, 256, 264
Mount Wilson Observatory, 86

National Research Council, 68, 107
National Socialist German Workers'
Party, see Nazi party
Nazi party, xi, 85, 87, 94, 103, 108,
126, 144, 147, 150, 151,
165
New York University, 168, 172, 185,
196
Newark, 32, 34, 36–41, 45, 52, 67,
70, 76
Newcomb, Simon, 11
Newton, Hubert, 20
Newton, Isaac, 84, 85
Noether, Emmy, 119
Northwestern University, 19, 20

Olden Farm, 191
Oppenheimer, J. Robert, 84, 142,
143, 252–259
Osgood, William, 11, 20, 62, 157
Osler, William, 22

Pais, Abraham, 254, 256, 258
Panic of 1857, 16
Panic of 1873, 2, 16, 73
Panic of 1893, 12, 32, 73
Panofsky, Erwin, 174, 175, 180, 184,
189, 190, 193, 194, 196,
200, 202, 203, 224, 232,
246, 253, 255
hiring, 168–172, 184–186
Pauli, Wolfgang, 182, 189, 190, 197,
248–249, 251

Pauling, Linus, 252
PhD program, *see* Institute for
        Advanced Study,
        postdoctoral program,
        evolution of
Phelps, Celeste, 156
Pillsbury, George, 22
Planck, Max, 83, 84
Poincaré Conjecture, 262
Poincaré's Geometric Theorem, 63
Poincaré, Henri, 61, 62, 65, 105, 128
preceptor, 61, 104, 113
Princeton, 1, 59, 66, 75, 97, 102, 154,
        189
Princeton University, 21, 45, 57, 58,
        61, 62, 77, 106, 191, 216,
        248
privatdozent, 83, 91, 117, 139, 147,
        154
Prussian Education Ministry, 109, 126

Reed, Walter, 9
Reichstag, 87, 144
Richardson, Elizabeth, *see* Veblen,
        Elizabeth Richardson
Richardson, R. G. D., 98, 100
Riefler, Winfield, 178–184, 187, 189,
        190, 194, 195, 198–199,
        202, 203, 205–207,
        210–212, 221, 223, 232,
        254
Riemann, Bernard, 1, 11, 93
Rockefeller Foundation, 186, 199,
        205, 208, 209, 212, 218,
        227, 228, 230, 234, 241,
        242
Rockefeller Institute for Medical
        Research, 24–26, 37, 49
Rockefeller, Eliza, 15
Rockefeller, John D., Jr., 23–25
Rockefeller, John D., Sr., 15–29, 31,
        35, 37, 44, 48, 95

Rockefeller, William, 15
Rome, 64
Roosevelt, Franklin, 122, 144, 151,
        173, 237
Rose, Wickliffe, 29
Rosenwald, Julius, 28
Rowland, Henry, 6, 53, 55, 187

Sabin, Florence, 49, 56
Samuelson, Paul, 64
Schaap, Michael, 247
Schrödinger, Anny, 154
Schrödinger, Erwin, 139, 154–155,
        204
Schumpeter, Joseph, 66, 74, 76, 176
Selberg, Atle, 254, 256, 260, 264
Siegel, Carl Ludwig, 157, 248–249,
        251
Smale, Stephen, 262
Smith, David Eugene, 56, 59
Smith, Theobald, 24
South Orange, 34, 35
Standard Oil Company, 15, 16
Stern, Beatrice, 143
Stewart, Walter, 162–163, 166, 167,
        176–179, 181, 184, 198,
        215, 217, 221, 222, 227,
        229, 231–233, 239, 247,
        249, 254
        hiring, 166, 167, 172–173,
        206–212
Stigler, George, 64
Story, William, 132
Straus, Percy, 49, 75, 185–186,
        240–241
Strauss, Lewis, 251, 252, 254–255
Stromgren, Bengt, 256
Swarthmore College, 50, 55, 223,
        240–242
Sylvester, James Joseph, 5–7, 9–12,
        19, 21, 53, 55, 60, 134, 186
Szilard, Leo, 236–237

Thompson, Homer, 251, 253
Thomson, J. J., 85
Truman, Harry, 250, 252, 255

Uhlenbeck, George, 253
University of California, 3, 4
University of Chicago, 17–22, 24, 25,
        28, 37, 48, 59, 62, 63, 184,
        250
University of Iowa Medical School, 29
University of Pennsylvania, 250
University of Rochester, 28, 37
University of Wisconsin, 3, 61

van Kampen, Egbertus, 153
Vanderslice, John, 116
Veblen, Elizabeth Richardson, 105
Veblen, Oswald, x, xi, 21, 59, 61, 65,
        75, 77–78, 89, 94,
        102–116, 125–128, 130,
        133, 134, 145, 146, 155,
        157, 160, 169, 175, 177,
        178, 181, 183, 186, 187,
        189–192, 196, 197, 202,
        206–208, 211, 213, 215,
        216, 218, 221, 223–225,
        227, 228, 230–231, 234,
        236, 237, 239–240,
        243–244, 246–248, 256,
        264
    attempt to start mathematics
        institute, 67–69, 107
    biography, 103–107
    identifies faculty, 77, 119–121,
        123, 129, 131, 134–137,
        139, 144, 156, 205, 249
    shapes postdoctoral program,
        121–122, 152–154,
        203–204, 261–262
Veblen, Thorstein, 103
Vietoris, Leopold, 117
Viner, Jacob, 63–66, 75, 79, 94, 159,
        160, 166, 180, 187, 248

Voevodsky, Vladimir, 264
von Hindenburg, Paul, 108, 110
von Neumann, John, x, 1, 115, 142,
        146, 155, 157, 176, 187,
        189, 194, 197, 208, 232,
        249, 256, 263, 264
    biography, 137–141
    computer project, 249–251
    hiring, 135–137
von Papen, Franz, 110

Walsh, Joseph, 157
Warburg, Paul, 57–58, 73
Warren, Robert, 210–213, 215, 217,
        219, 221, 222, 232, 247,
        254
Washington University, 37
Washington, DC, 57, 75, 76
Weed, Lewis, 49, 56, 76, 183,
        223–224, 226, 227, 245,
        248
Weil, André, 256, 258, 259, 264
Weizmann, Chaim, 85
Welch, William, 7–9, 24–27, 37, 53,
        55, 71, 186
Wellesley College, 197
Weyl, Hella Joseph, 92, 94
Weyl, Hermann, 59, 66, 89, 94, 95,
        98, 107, 114, 119, 127,
        134, 139, 154–157, 161,
        178, 187, 189, 197, 211,
        224, 232, 249, 256, 264
    biography, 90–94
    breakdown, 136–137, 145
    decision difficulty, 93–94,
        102–103, 108, 109, 112,
        114–116, 125–127,
        134–137
    hiring, 77, 89, 90, 96, 102–103,
        109, 112, 114–116, 130,
        145–148, 163

White, Henry Seely, 20
White, Morton, 259
Whitney, Hassler, 122, 256, 264
Wiener, Norbert, 119, 144, 156
Wigner, Eugene, 138, 140, 155
Wilson, Woodrow, 6, 61, 122
Witten, Edward, 264
Wolfensohn, James, 260

Wolman, Leo, 210
Woolf, Harry, 260

Yale University, 3, 4, 17, 20, 162
    Scientific School, 3
Yang, Chen Ning, 256, 258
Yukawa, Hideki, 253

Zariski, Oscar, 153, 157